AF576003

Environmental Compatible Circuit Breaker Technologies

Environmental Compatible Circuit Breaker Technologies

Editor

Dirk Uhrlandt

MDPI • Basel • Beijing • Wuhan • Barcelona • Belgrade • Manchester • Tokyo • Cluj • Tianjin

Editor
Dirk Uhrlandt
Leibniz Institute for Plasma
Science and Technology
Germany
University of Rostock
Germany

Editorial Office
MDPI
St. Alban-Anlage 66
4052 Basel, Switzerland

This is a reprint of articles from the Special Issue published online in the open access journal *Energies* (ISSN 1996-1073) (available at: https://www.mdpi.com/journal/energies/special_issues/circuit_breaker).

For citation purposes, cite each article independently as indicated on the article page online and as indicated below:

LastName, A.A.; LastName, B.B.; LastName, C.C. Article Title. *Journal Name* **Year**, *Volume Number*, Page Range.

ISBN 978-3-0365-0384-4 (Hbk)
ISBN 978-3-0365-0385-1 (PDF)

Contents

About the Editor . . . . **vii**

Preface to "Environmental Compatible Circuit Breaker Technologies" . . . . **ix**

Odd Christian Feet, Martin Seeger, Daniel Over, Kaveh Niayesh and Frank Mauseth
Breakdown at Multiple Protrusions in SF_6 and CO_2
Reprinted from: *Energies* , *13*, 4449, doi:10.3390/en13174449 . . . . **1**

Ralf Methling, Alireza Khakpour, Nicolas Götte, and Dirk Uhrlandt
Ablation-Dominated Arcs in CO_2 Atmosphere—Part I: Temperature Determination near Current Zero
Reprinted from: *Energies* , *13*, 4714, doi:10.3390/en13184714 . . . . **21**

Ralf Methling, Nicolas Götte, and Dirk Uhrlandt
Ablation-Dominated Arcs in CO_2 Atmosphere—Part II: Molecule Emission and Absorption
Reprinted from: *Energies* , *13*, 4720, doi:10.3390/en13184720 . . . . **41**

Ali Kadivar and Kaveh Niayesh
Effects of Fast Elongation on Switching Arcs Characteristics in Fast Air Switches
Reprinted from: *Energies* , *13*, 4846, doi:10.3390/en13184846 . . . . **59**

Sergey Gortschakow, Steffen Franke, Ralf Methling, Diego Gonzalez, Andreas Lawall, Erik D. Taylor and Frank Graskowski
Properties of Vacuum Arcs Generated by Switching RMF Contacts at Different Ignition Positions
Reprinted from: *Energies* , *13*, 5596, doi:10.3390/en13215596 . . . . **83**

Ehsan Hashemi and Kaveh Niayesh
DC Current Interruption Based on Vacuum Arc Impacted by Ultra-Fast Transverse Magnetic Field
Reprinted from: *Energies* , *13*, 4644, doi:10.3390/en13184644 . . . . **97**

Dequan Wang, Minfu Liao, Rufan Wang, Tenghui Li, Jun Qiu, Jinjin Li, Xiongying Duan and Jiyan Zou
Research on Vacuum Arc Commutation Characteristics of a Natural-Commutate Hybrid DC Circuit Breaker
Reprinted from: *Energies* , *13*, 4823, doi:10.3390/en13184823 . . . . **111**

Yun Geng, Xiaofei Yao, Jinlong Dong, Xue Liu, Yingsan Geng, Zhiyuan Liu, Jing Peng and Ke Wang
Experimental Investigation of the Prestrike Characteristics of a Double-Break Vacuum Circuit Breaker under DC Voltages
Reprinted from: *Energies* , *13*, 3217, doi:10.3390/en13123217 . . . . **127**

About the Editor

Dirk Uhrlandt has been with the Leibniz Institute for Plasma Science and Technology e.V. (INP Greifswald) in Germany since 1993. At present, he is the head of the research division "Materials and Energy" and a scientific board member of the INP. He is also a full professor of high current and high voltage engineering at the Faculty for Computer Science and Electrical Engineering of the University of Rostock, Germany, since 2017. He received his Ph.D. in theoretical physics from Greifswald University, Germany, in 1997. His main research interests are the physics and application of thermal plasmas. This includes arc behavior and arc–electrode interaction in vacuum interrupters, in high-voltage circuit breakers and arrestors as well as in low-voltage switchgear. He is also interested in arc behavior and material transfer in arc welding processes. His work comprises optical diagnostics including emissions and absorption spectroscopy, as well as the magneto-hydrodynamic simulation and non-equilibrium modeling of arcs.

Preface to "Environmental Compatible Circuit Breaker Technologies"

Changing demands on power transmission and distribution grids have initiated comprehensive research and innovations in the field of circuit breaker technology for high and medium voltages. General trends toward higher voltage levels and increased reliability are supplemented by challenges of more distributed power generation and the need for large distance transmission due to the increasing use of renewable energies. An additional demand is the environmental compatibility of the components. Circuit breakers as a key component of electric grids have to safely cut off not only the normal load current but also the high fault currents caused by short circuits or ground faults on the load side.

Recent research and development have focused on attractive topics like the replacement of environmentally harmful SF_6 in gas breakers and new solutions for high-voltage direct current (DC) switching, which is required for large distance transmission and the connection of offshore wind farms for example. The expansion of the application range of vacuum switchgear to higher voltage levels offers alternatives for SF_6 gas breakers. Vacuum switchgear is also an interesting option for hybrid circuit breaking concepts. Despite extensive research recently, satisfactory and feasible solutions are largely missing.

This Special Issue comprises eight peer-reviewed papers, which address recent findings in the field of environment compatible switching and circuit breaker technologies. The first three papers deal with the issue of the replacement of SF_6 by CO_2 in gaseous insulated switchgear. The first paper considers the gaseous insulation and the experimental observation of the electrical breakdown in contact gaps filled with CO_2 in comparison with SF_6 under the impact of the occurrence of protrusions at the electrodes. The second and third papers concern ablation dominated arcs in a circuit breaker operated in CO_2 focussing on the experimental study of plasma properties around the current zero-crossing, as well as focussing on the spectroscopy of molecule radiation and their potential for the further study of such arcs and the ablation process. A modeling study of ultra-fast switches in air with relevance for hybrid fault current limiters and hybrid high-voltage DC interrupters is presented in the fourth paper. It focuses on the high-speed elongation of the arc to reach rapid voltage increases.

The next four papers deal with vacuum circuit breakers as an environmentally compatible technology, where the first paper in this series presents fundamental studies of the vacuum arc for radial magnetic field contacts and the dependence of the arc properties and post-arc parameters on the arc ignition point. The next three papers consider the challenge of DC current interruption based on vacuum arcs. The sixth paper presents a study of the idea to create a zero-crossing of the current by applying an external ultra-fast transverse magnetic field to the vacuum arc in parallel to a capacitor. A hybrid DC circuit breaker is studied in the seventh paper, focussing on the vacuum arc commutation and its dependence on the switching parameters. The eighth paper considers double-break vacuum circuit breakers as an option for handling higher DC voltages and reports on experimental studies of the prestrike probability characteristics.

Dirk Uhrlandt

Editor

Article

Breakdown at Multiple Protrusions in SF_6 and CO_2

Odd Christian Feet [1], Martin Seeger [2,*], Daniel Over [2], Kaveh Niayesh [3] and Frank Mauseth [3]

1 ABB Motion Norway, 0666 Oslo, Norway; odd-christian.feet@no.abb.com
2 Hitachi ABB Power Grids Research, 5401 Baden-Dättwil, Switzerland; daniel.over@hitachi-powergrids.com
3 Department of Electric Power Engineering, Norwegian University of Science and Technology, 7491 Trondheim, Norway; kaveh.niayesh@ntnu.no (K.N.); frank.mauseth@ntnu.no (F.M.)
* Correspondence: martin.seeger@hitachi-powergrids.com

Received: 7 July 2020; Accepted: 19 August 2020; Published: 27 August 2020

Abstract: The electric breakdown at single and multiple protrusions in SF_6 and CO_2 is investigated at 0.4 and 0.6 MPa, respectively. Additionally, the breakdown fields at rough surfaces of two different areas were determined. From the measurements, breakdown probability distributions for single protrusions were determined and fitted by Weibull distributions. This allowed the determination of statistical enlargement laws for the 50% breakdown probability fields E_{50}. Such enlargement laws describe, for example, the scaling of breakdown field with electrode area or number of protrusions. The predictions were compared to the experimental data, and both agreement and discrepancies were observed depending on polarity and number of protrusions and gas. Discharge predictions including first electron, streamer inception and crossing, as well as leader propagation, gave further insight to this. It was found that predictions from enlargement laws based on statistical processes may not describe the measured breakdown fields well and that relevant physical breakdown criteria must also be considered.

Keywords: gaseous breakdown; SF_6; CO_2; surface roughness; statistical enlargement laws

---

1. Introduction

Gaseous insulation is used in many high voltage applications as, for example, gas-insulated switchgear (GIS) and circuit breakers (CB) [1–6]. For compact insulation compressed gas in the pressure range of 0.13–1 MPa is used. In most applications today, SF_6 is used; however, it is a strong greenhouse gas (e.g., [7,8]). Therefore, the search for alternative insulation gases has significantly increased during the last years. The most promising gas in HV switchgear applications for replacing SF_6 is CO_2, which is used in mixtures with low-additive concentrations of, e.g., O_2, Perfluoroketones (PFK) or Perfluoronitriles (PFN) [8–12]. The additive concentrations are typically in the range of a few percent due to boiling point requirements. This will be referred to as CO_2-based insulation.

Gaseous insulation strength depends on the critical field of the gas [2,5,6], the field at which ionization leads to multiplication of electrons within an avalanche. This is determined by the zero crossing of the effective ionization coefficient, which is specific for a gas. The slope of the effective ionization coefficient determines the sensitivity of the insulation system to the surface imperfections [13–24]. Thus, depending on the properties of the gases, the insulation performance is determined not only by the critical field of the gas but also by imperfections like surface roughness or particulate contamination. This is especially more pronounced at higher pressures of 0.7 to 1 MPa, which is the preferred pressure range for CO_2-based insulation [25–33]. Since CO_2 is the base gas in these mixtures, a good understanding of the insulation characteristics of CO_2 at conditions of practical applications is needed.

Several investigations addressed the effect of electrode surface roughness and representations of those by single or multiple protrusions in SF_6—see [14,16–18,21]. Single-protrusion models have shown reasonable agreement with measurements when taking into account the availability of a start electron for an avalanche, streamer inception and leader propagation [34–38]. Surface roughness and the effect of insulating coatings in SF_6 have been investigated experimentally by [23,24].

CO_2-based insulation is still less investigated than SF_6, and investigations with fluorinated additives to CO_2 just recently appeared. The effect of surface roughness and protrusions in pure CO_2 was investigated by [29–33,36,37], respectively. Breakdown at surfaces in CO_2-PFK and CO_2-PFN mixtures is reviewed in [8–10]. The effect of surface roughness at pressures above 1 MPa was investigated, e.g., by [27,36]. Breakdown at single protrusions in CO_2 was investigated by [37], showing that streamer propagation, in contrast to SF_6, also plays an important role for the breakdown in practical configurations, like in the presence of particulate contamination.

Several investigations addressed the effect of enlargement laws, i.e., the reduction of the breakdown field by the size of the electrode area or volume and time laws—see [5,6,13,39]. A theoretical investigation on approximation of surface roughness by an array of multiple protrusions was done by [40]. In this investigation, the surface roughness was approximated by multiple protrusions of given length and with hemispherical tip. The influence of the distance between protrusions was also addressed. Application of discharge inception and breakdown criteria based on start electron, streamer inception and leader inception were used to estimate breakdown fields for such arrays.

The previous investigations have shown that there is still only limited understanding how to describe the surface roughness induced breakdown fields by single and multiple protrusions, especially when considering the surface area and time scaling effects. Therefore, measurements are still needed to characterize breakdown fields in practical applications as, e.g., shown in [5,27,29,34]. It would be desirable to have models for surface roughness or protrusion/particle induced breakdown which can, for arbitrary gases, electrode size and number of protrusions, describe the breakdown fields sufficiently reliable. The present investigation addresses this in a first step by measuring breakdown fields at single and multiple protrusions for short duration voltage waveform, which is similar to Lightning Impulse (LI) in standard testing according to IEC [6]. This procedure is similar to the theoretical study of [40]. Investigation of different gases in this study allows for understanding the differences between strongly attaching (SF_6) and weakly attaching (CO_2) gases. A novel test setup with a multiple protrusion array in a uniform background field is used and fill pressures of 0.4 MPa SF_6 and 0.6 MPa CO_2 are tested. The protrusion lengths are varied between 50 μm and 2 mm. The number of protrusions is varied between 1 and 100. Protrusions of small lengths of a few 10 to few 100 μm correspond to the surface roughness, whereas long protrusions of more than 500 μm correspond to particulate contamination or severe surface damages [35]. The effect of the number of protrusions, i.e., the statistical enlargement law, is addressed by scaling of the measured breakdown probability distributions using Weibull approximations—see [6,39,41,42]. This allows the validity of enlargement laws for such well-defined situations to be checked. Further insight is obtained by comparison to theoretical discharge inception and breakdown models based on start electron, streamer inception and streamer and leader propagation from [34–38]. Additionally, the surface area scaling is investigated experimentally on rough electrodes of similar surface structure. Results are interpreted with the findings from the protrusion arrays.

Section 2 shows the experimental setups and Section 3 presents the results. Discussion and conclusion are given in Section 4.

2. Test Setups

2.1. Test Circuit and Protrusion Setup

The experiments and the analysis were performed at Hitachi ABB Power Grids Research, Switzerland. The setup is shown schematically in Figure 1. It is a similar setup as used in previous investigations where the partial discharges and breakdown at a single protrusion was investigated [37,38]. In the present investigation, the plate electrode with the protrusion on the ground side was replaced by a plate with a protrusion array. The protrusion array plate had 20 × 20 holes of 1.2 mm diameter and 2 mm center to center spacing—see Figure 2a. The protrusions were at ground potential. When we refer to polarity in the following always the polarity of the protrusions is meant.

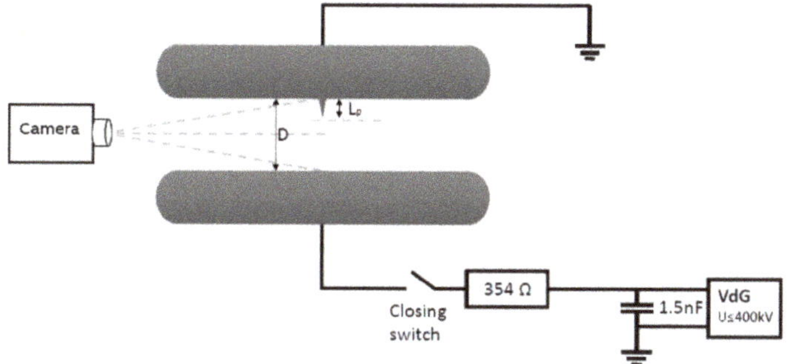

Figure 1. Schematic setup with a single protrusion. The gap length was D = 10 or 15 mm. The load capacitance was charged via a Van de Graaf (VdG) DC source and switched via a resistor and closing switch onto the test gap. The discharges were observed by a high-resolution digital camera (Nikon D9000).

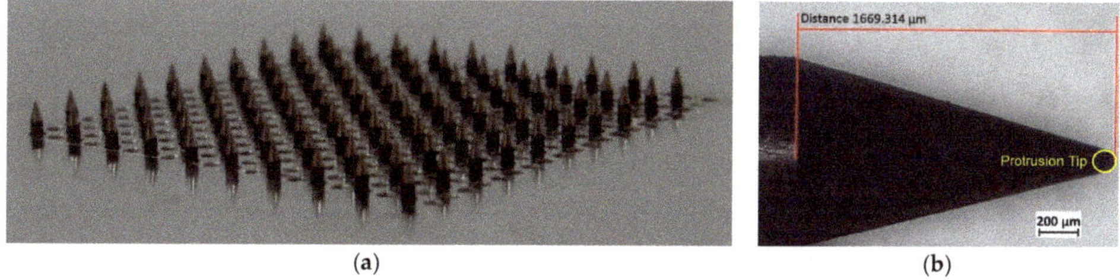

(**a**) (**b**)

Figure 2. Protrusion array. (**a**) Protrusion array with 100 protrusions. (**b**) Protrusion tip details.

In the shown case in Figure 2a, 100 protrusions were inserted, but also fewer protrusions were tested. The protrusions were fixed to a matching grid by a magnet on a table below the perforated plate. The matching grid was moved by a stepper motor (Thorlabs GmbH, Bergkirchen, Germany, Model ZFS13B) with high positioning accuracy (<1 µm). The setup was placed, as in previous experiments, in a GIS test vessel filled with 0.4 MPa SF_6 or 0.6 MPa CO_2. With the stepper motor the protrusion lengths could be adjusted from outside, i.e., without opening the GIS. The overall reproducibility of the length adjustment was experimentally determined to be better than 20 µm for a single protrusion. In experiments with several protrusions, the total variation of lengths was less than 100 µm.

The protrusions were made out of steel with diameter 1 mm and had a conical tip of 1.6–1.7 mm length and tip radius of 66 µm at its end, marked by the yellow circle—see Figure 2b. The spacing between the protrusions was usually 4 mm. Only for CO_2 when using 20 protrusions also a spacing of 2 mm was tested. The protrusions were replaced regularly, typically after a few hundred tests. The effect of erosion was checked by repeating experiments in new and worn conditions. The differences in the breakdown voltages due to this were within the uncertainties of protrusion lengths adjustments, i.e., repeating experiments with the same protrusion in new condition led to a similar breakdown voltage as in worn condition.

The applied voltage was a stepped DC pulse, as used in [37,38], with a rise time of 200 ns and voltage application time of 15 s, typically. In experiments with multiple protrusions, the voltage dropped due to partial discharges after a few 10 µs; thus, we focused on early breakdowns in the time range from the voltage peak to 10 μs after the peak was reached. From the time to breakdown, this can be associated with standard lightning impulse (LI) breakdown.

The applied voltage peak was varied over a wide range in order to obtain breakdown and withstand values. Typically, at least 100 tests were done per case. Between the tests, a small bias voltage of 20 kV was applied for 2 s to remove remaining ions from the gap, which could influence the start electron statistics. Thus, in the present investigation we determine breakdown fields at protrusions which can be related to LI standard wave shape. For small protrusion size in the order of 100 μm, breakdown also occurred at the plate electrode, e.g., at dust particles. With a single-shot digital camera (see Figure 1), it was checked that only breakdowns that occurred at a protrusion were taken into consideration.

2.2. Electrodes with Surface Roughness

Two different electrode types were tested: a small plug-type contact of 19 mm diameter, as seen in Figure 3, and a larger Rogowski-shaped electrode of 120 mm diameter. The electrodes were placed against a plate electrode, as in setup 1 with a 10 mm gap. Both electrodes were made from stainless steel and were sandblasted with Corundum white grains, resulting in a surface roughness of R_a = 5–7 μm and R_z = 62–65 μm with the average roughness $R_a = \frac{1}{L}\int_0^L |z(x)| \cdot dx$ over the sampling length L of the contour $z(x)$ and the mean peak-to-valley height $R_z = \frac{1}{N} \cdot \sum_{i=1}^{N} R_{t,i}$ over an assessment length consisting of various sampling sections with the maximum peak to valley height $R_{t,i}$ within each section; see [43] for details. The surface roughness was determined by a non-contacting 3D profilometer Hyperion Compact from OPM. The effective area of the electrodes exposed to the electric field was about 240 and 6450 mm^2, respectively. This effective field exposed area was determined from the location of the breakdown marks after test, as can be, e.g., judged from the photographs in Figure 3 for the plug-type contact. For the plug-type contact, the field enhancement factor was calculated to be 1.42.

Figure 3. Plug-type contact with diameter 19 and 10 mm length in new (**right**) and worn condition after test (**left**). The surface has a biradial shape with a radius of 5 mm at the side and 20 mm at the tip, respectively.

3. Evaluation Procedures

3.1. Breakdown Probability Distributions

From the recorded breakdown fields, the empirical cumulative breakdown probability distributions were determined by the Turnbull algorithm [44] and fitted with three-parameter Weibull distributions [41] as proposed in [6,42]. The zero crossing E_0 of the Weibull distribution was set to the streamer inception field, calculated for each protrusion length as will be described below in Section 3.2.2. From the distributions, the 50% BD probability field E_{50} and the standard deviation σ was determined. Note that these values refer to the background field, i.e., E_{50} = U_{50}/D with 50% breakdown probability voltage U_{50} and the gap length D. Generally, for the identical protrusions of m, the enlargement law based on the Weibull distributions generally follows from [6]

$$p(E) = 1 - e^{-m \cdot [\frac{E-E_0}{E_{63}-E_0}]^{\gamma}} \quad (1)$$

With the shape parameter γ and the 63% breakdown probability field E_{63}. In the present case for γ and E_{63} the values from the fits to the empirical cumulative breakdown probability distributions were used. An example of the enlargement scaling is shown in Figure 4. With increasing number of protrusions, the breakdown probability distribution becomes steeper and shifts towards E_0. Note, that such scaling is purely a statistical process and does not necessarily describe the scaling of any physical processes. This can be interpreted as follows: if the breakdown probability distribution follows the scaling from (1), it can be expected that the underlying processes are of statistical nature, e.g., by the availability of a first electron to start an avalanche. For electrodes with surface roughness, m is the ratio of the surface areas.

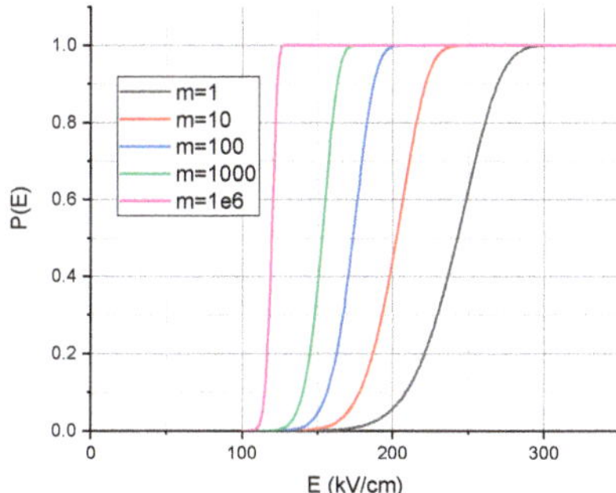

Figure 4. Example of Weibull distributions for increasing number of protrusions m. In this example, E_{63} = 250 kV/cm, E_0 = 100 kV/cm, γ = 7 was chosen arbitrarily.

3.2. Discharge Inception and Breakdown Models

In the following subsections, models for the description of first electron, streamer inception and leader propagation will be described. The models are the same as presented in [38] for SF_6 and will only be briefly summarized here with the specific adaptions for CO_2. For these models, the decay of the electric field at the protrusion and at the electrode surface is needed. The field enhancement in the axis of symmetry in front of a protrusion was calculated by a multipole approximation

$$\frac{E(x)}{E_b} \approx 1 + \frac{l/r - 1}{(x/r+1)^2} + \frac{2}{(x/r+1)^3} \quad (2)$$

with the protrusion length and radius l and r, respectively, as used in [36]. E_b is the undisturbed background field. It was checked by 3D electric field calculations with COMSOL for some cases that this approximation was sufficiently precise for the present purpose. With r = 100 µm as assumed protrusion radius, average deviations in the field decay were within 10%. The formula is only correct until $l = r$, i.e., l = 100 µm. Therefore, predictions will only be shown for 100 µm length and higher. For rough surfaces $l/r = 1$ with l = 20 µm as a typical peak height of roughness structures was assumed. This is similar to the approach of [14]. The chosen length is deduced from the measured R_z by assuming $l = R_z/3$. Note that R_z describes the average difference between highest peaks and lowest valleys of the surface.

3.2.1. First Electron

The electric field needed for a start electron at positive polarity in SF_6 can be estimated from the volume where the electric field is above the critical field and where an electron can be detached from a negative ion in the available time of voltage application, i.e., in the present case within 10 µs—see [38] for details of the model. All parameters for SF_6 were taken from [34].

At negative polarity, a start electron is assumed to be delivered from the electrode surface. For SF_6, this could be described reasonably well by using the Fowler–Nordheim (FN) equation in [34], which gives the electron production rate emitted from a surface area at given electric field. The statistical time lag at negative polarity is then

$$t_s = \left(A_{eff} \cdot 10^{4.52/\sqrt{\Phi}} \cdot 1.54 \cdot 10^{-6} \cdot \frac{(\beta \cdot E_b)}{\Phi} \cdot \exp\left[-\left(\Phi^{1.5} \cdot 2.84 \cdot 10^9\right)/(\beta \cdot E_b)\right]/e\right)^{-1} \quad (3)$$

where $\Phi \approx 4.5$ eV is the work function for steel, e the elementary charge, $\beta = (2 + l/r) \cdot \beta_2$ a field enhancement factor at to the protrusion tip (see also (2)) and due to micro-surface roughness and A_{eff} the effective electron emitting area. For SF_6 values for β_2 and A_{eff} were taken from the fits in [34] with $\beta_2 = 20$ and $A_{eff} = 10^{-16}$ m^2. From this, the necessary electric field for a start electron within the available time can be deduced.

For CO_2, the same approach and models as for SF_6 were used but with CO_2 specific adaptions and empirical approximations for the relevant parameters. For the statistical time lags, the field dependence of the electron detachment rate, the equilibrium negative ion concentration and the parameters in the FN equation were adapted to experimental data from [37]. At positive polarity, best agreement, within a large scatter, was achieved by empirically using an equilibrium concentration of 10^4 ions/m^3 at 0.1 MPa in the pressure-dependent ion concentration, as given in [38], and a field dependence of the electron detachment rate coefficient of

$$\delta = 10 \cdot \left(\frac{E}{E_{cr}}\right)^{13} \; [1/s] \quad (4)$$

with the critical electric field of CO_2 at pressure p: $E_{cr} = p \cdot (E/p)_{cr,0}$, with $(E/p)_{cr,0} = 23$ V/(m·Pa) [37]. To our knowledge, no experimental data are available to verify this scaling of the electron detachment rate. At negative polarity, the field enhancement parameter β_2 in (3) was set to $\beta_2 = 70$ and the effective area needed to be set to $A_{eff} = 10^{-24}$ m^2 to achieve best agreement with the experimental data. Such small value for A_{eff} is probably not realistic, but one has to consider the simplicity of the FN equation. The approximations have to be seen, therefore, just as physically motivated fits to the experimental data which are valid in the given parameter range of protrusions up to a 4.5 mm length. Examples for statistical time lags for negative and positive polarity with the data from [37] are shown in Figure 5. For surface roughness, values from the rod plane experiment in [37] are used. The field enhancement factor at the surface of the rod (=$E_{surface}$/(1/D)) of the rod in these experiments was about 17. The background field in the plots is the undisturbed electric field U/D without field enhancement by a protrusion or by the rod.

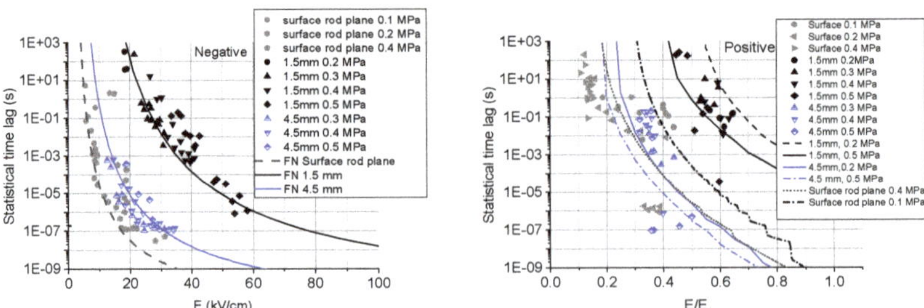

Figure 5. Statistical time lags in CO_2 at negative (**left**) and positive (**right**) polarity vs. background electric field for technical surface roughness and protrusions with 1.5 and 4.5 mm lengths. The curves are the model predictions and the symbols result from measurements given in [37]. For the technical surface roughness estimates, a 20-μm hemispherical protrusion on the surface was assumed [36]. Note that, for positive polarity, the electric field is normalized to the critical field E_{cr}, which yields a less scattered representation for the different pressures used.

3.2.2. Streamer Inception

Streamer inception fields were calculated using the semi-empirical approach in [38]

$$K = \int_0^{l_{cr}} \overline{\alpha} \cdot dz \quad (5)$$

With the ionization integral parameter $K = 10.5$ and 13 for SF_6 [45] and CO_2 [37], respectively. The effective ionization coefficient $\overline{\alpha}$ was taken from [38] for SF_6 and [37] for CO_2. The integration is done along the axis of symmetry starting from the protrusion tip at $z = 0$ where the electric field is above the critical field, up to the distance $z = l_{cr}$, where the field has dropped to the critical field.

3.2.3. Streamer Crossing and Spark Transition

Streamer crossing followed by spark transition is only relevant for CO_2, since in SF_6 this occurs at the critical field, which is much higher than leader propagation fields [38]. In [37], it was shown that streamer crossing for long duration voltage application times of several 100 μs is approximately sufficient for breakdown in the pressure range investigated here. In case of short-duration voltage application time, a higher field than that necessary for streamer crossing is needed to allow for the spark transition within the short available time of 10 μs, since the heating processes in the streamer channel need a sufficient electric field and time. This field increase was estimated from [46] to roughly 35% compared to the streamer crossing field, which is based on the breakdown voltage ratio for short and long duration waveforms.

3.2.4. Leader Propagation and Breakdown

As shown in [35,38], leader propagation through the gap can be associated with breakdown. Leader inception and propagation was calculated using the previous model from [35] for SF_6. In this model, the streamer corona charge at the protrusion tip or a propagating leader is calculated and fed into the streamer or leader channel. This leads to stepped heating followed by stepwise leader propagation through the gap. Crossing 90% of the gap distance was defined as sufficient for breakdown. For CO_2, the same model but with adapted thermodynamic properties of mass density, enthalpy, velocity of sound and effective ionization coefficients [47] from our own, in-house-developed solver were used.

4. Results

4.1. Protrusions with SF_6

The results for the single protrusions in SF_6 are shown in Figure 6 for positive and negative polarity. The E_{50} background fields are normalized to the critical field, which eases the interpretations and comparison between different gases. In agreement with [38], the breakdown values for positive polarity are slightly lower than those at negative polarity. Only for protrusion length of 200 μm or less the differences between the polarities are within the scatter. At such small lengths, many breakdowns also happened away from the protrusions, which were, however, not considered for the determination of E_{50}, as mentioned above. The decrease of the normalized E_{50} breakdown field with increasing protrusion length is reasonably described by the prediction for leader breakdown, which occurs significantly below the critical field. Streamer inception and the first electron criterion are fulfilled at lower fields, such that leader breakdown is expected to be the decisive criterion. Only for very small lengths below 250 μm and positive polarity, the first electron criterion might require slightly higher fields than the leader breakdown. This is, however, within the uncertainties of the predictions. We can also observe a higher scatter for positive polarity compared to negative polarity.

Using 20 protrusions—see full symbols in Figure 7—lowers the normalized E_{50} breakdown fields significantly. For comparison, the experimental results and predictions for single protrusions are also shown. The difference between positive and negative polarity in the figure seems more pronounced than for single protrusions. The experimental breakdown fields are now lower than the predictions for leader breakdown for a single protrusion. The enlargement law predictions from (1) are shown by the black dash-dotted curves and agree for both polarities with the experimental results. Thus, the lowering of the experimental breakdown fields for 20 protrusions is well described by the enlargement law based on the breakdown probability distributions of single protrusions.

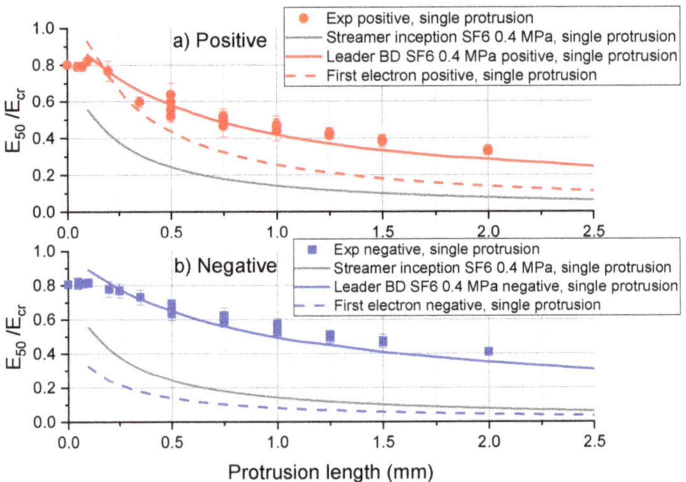

Figure 6. Normalized breakdown fields E_{50}/E_{cr} vs. protrusion length L for SF_6 and single protrusion at 0.4 MPa at (**a**) positive and (**b**) negative polarities.

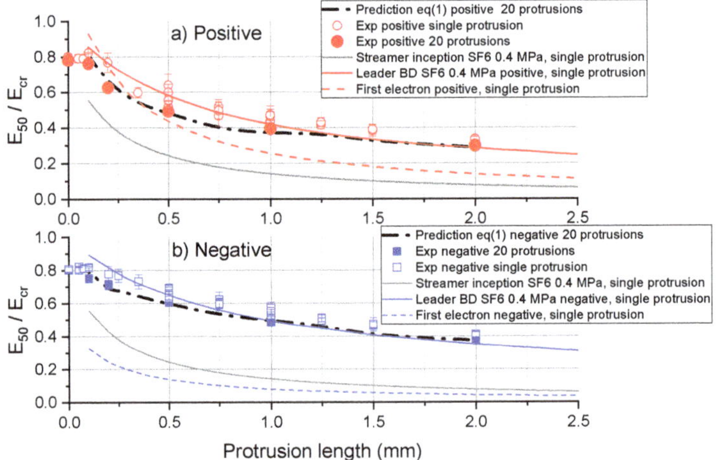

Figure 7. Normalized breakdown fields E_{50}/E_{cr} vs. protrusion length L for SF_6 and 20 protrusions. Not all protrusion lengths could be measured when using multiple protrusions at (**a**) positive and (**b**) negative polarities.

For 100 protrusions this is different at positive polarity—see Figure 8a. There is no further lowering of the breakdown fields when using 100 instead of 20 protrusions and the prediction of the enlargement law (1) is significantly lower than the measurement. There seems to be a lower limit which is not further exceeded when using a larger number of protrusions. This is not the case for negative polarity—see Figure 8b—where we see a good agreement of the enlargement law predictions with the experimental results for 100 protrusions.

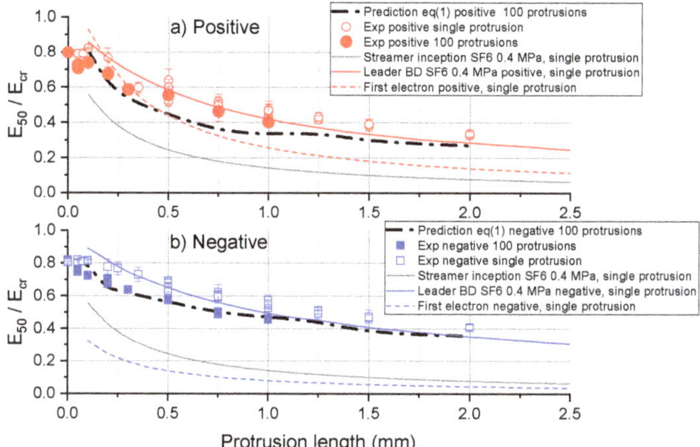

Figure 8. Normalized breakdown fields E_{50}/E_{cr} vs. protrusion length L for SF_6 and 100 protrusions. Not all protrusion lengths could be measured when using multiple protrusions at (**a**) positive and (**b**) negative polarities.

Typical partial discharge images, which were taken at breakdown events, are shown in Figure 9. It can be seen that many discharges happened in parallel, i.e., simultaneously within 10 µs. These discharges could be observed in case of breakdown since the discharge channels are illuminated by the strong light emission from the breakdown spark channel occurring elsewhere in the gap. These images are, therefore, similar to a Shadowgraphs or Schlieren images [48]. The discharge channels have the typical signature of leader channels [35], which confirms the interpretation from leader breakdown predictions, i.e., the breakdown is determined by leader propagation. The negative leader channels (Figure 9c) are thicker and less structured than positive leader channels (Figure 9b); this observation is in agreement with the expectation from [49].

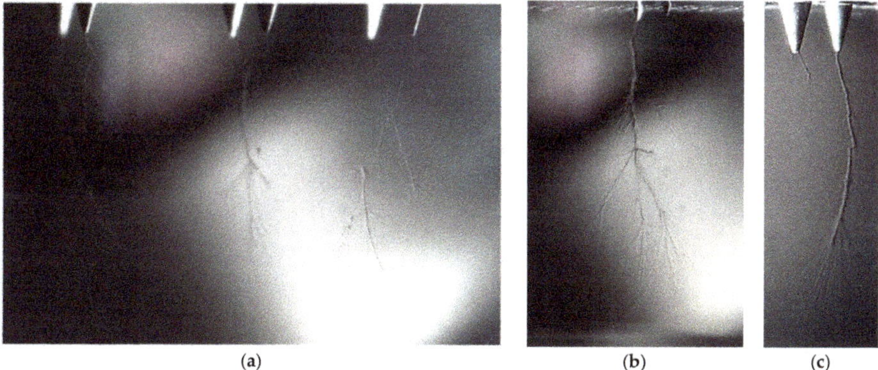

(**a**) (**b**) (**c**)

Figure 9. Images of partial discharges and breakdown in SF_6 at 0.4 MPa for (**a**) 2 mm length at positive polarity at 130 kV/cm, (**b**) 0.5 mm length at positive polarity at 188 kV/cm and (**c**) 1 mm length at negative polarity at 187 kV/cm. Note the different length scales of the images, which can be judged by the protrusion lengths.

4.2. Protrusions with CO_2

Results for single protrusions in CO_2 are shown in Figure 10. The figure shows the experimental normalized E_{50} breakdown values and predictions for first electron, streamer inception and leader breakdown and the background fields for streamer crossing and spark transition of negative streamers. The complete range where streamer crossing and spark transition is expected at negative polarity is indicated by the cyan colored area in the figure. For positive streamers we expect streamer crossing only at the critical field [37], which is, therefore, probably not relevant in the present case, since this needs fields higher than the leader breakdown criterion.

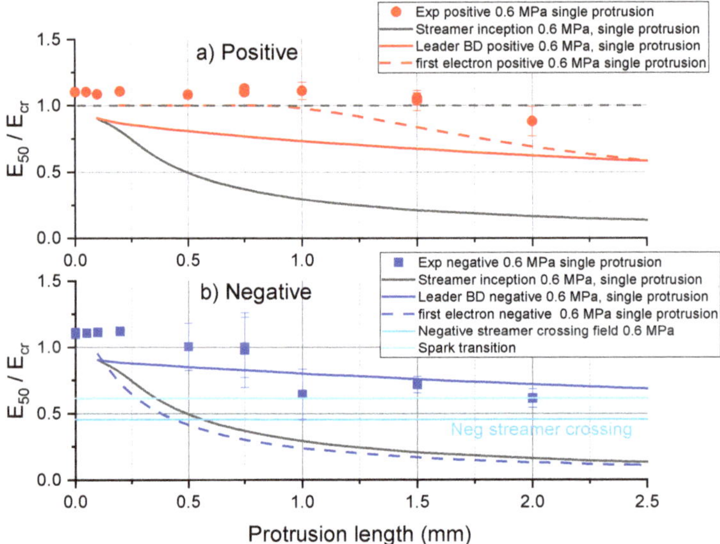

Figure 10. Normalized breakdown fields E_{50}/E_{cr} vs. protrusion length L for CO_2 and single protrusion at 0.6 MPa at (**a**) positive and (**b**) negative polarities.

For protrusion lengths below 1.5 mm with positive polarity, the breakdown happens at or slightly above the critical background field—see Figure 10a. The leader BD predictions do not explain this. Probably this can be explained by the first electron criterion (red dashed curve). At or above the critical field, avalanches can start everywhere in the gap, which is in agreement with the experiment. Many breakdowns did not occur at the protrusion but in other locations of the gap in this case. There was also a larger scatter in the breakdown fields, which is reflected in the error bars of the experimental breakdown values, especially at negative polarity.

At negative polarity, experimental E_{50} values are close to the leader breakdown predictions for protrusions larger than 1 mm—see Figure 10b.. An additional possibility for breakdown at negative polarity might be streamer crossing and transition to spark. Therefore, at negative polarity both mechanisms might be decisive for breakdown. From the partial discharge images—see Figure 11—this could not be decided unambiguously. For smaller protrusions, breakdown occurs at the critical field. As for positive polarity, breakdowns often occurred at the plate electrodes and not at the protrusion for smaller protrusion lengths.

Using 20 protrusions, see Figure 12, the measured breakdown fields decrease at both polarities compared to the results from the single protrusion experiments. For positive polarity, this is in agreement with the prediction from the enlargement law (1)—see Figure 12a. Note that predictions for the enlargement law were not possible for small protrusion lengths due to the limited amount of data and large uncertainties of the breakdown probability distributions. The measured breakdown values are for the positive polarity at the first electron prediction for single protrusions but still higher than the leader breakdown predictions. For negative polarity, a lowering of E_{50} values is observed mainly below 1 mm protrusion length—see Figure 12b. The measured breakdown fields below 0.5-mm protrusion length coincide with the first electron and streamer inception prediction. Interestingly, for protrusion lengths of 1 mm and more, no significant decrease of the E_{50} values for 20 protrusions compared to single protrusions is observed. The breakdown fields are at the expected spark transition level, which is assumed to be independent of protrusion length.

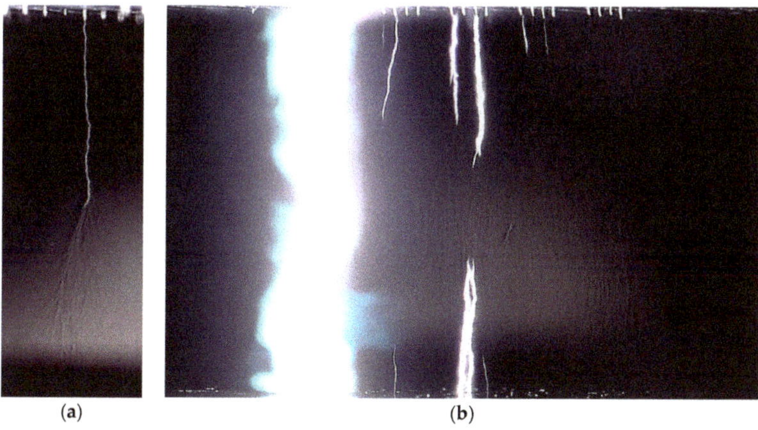

Figure 11. Images of partial discharges and breakdown in CO_2 at 0.6 MPa for (**a**) positive polarity at 115 kV/cm with 100-μm protrusions and (**b**) negative polarity at 139 kV/cm with 200-μm protrusions.

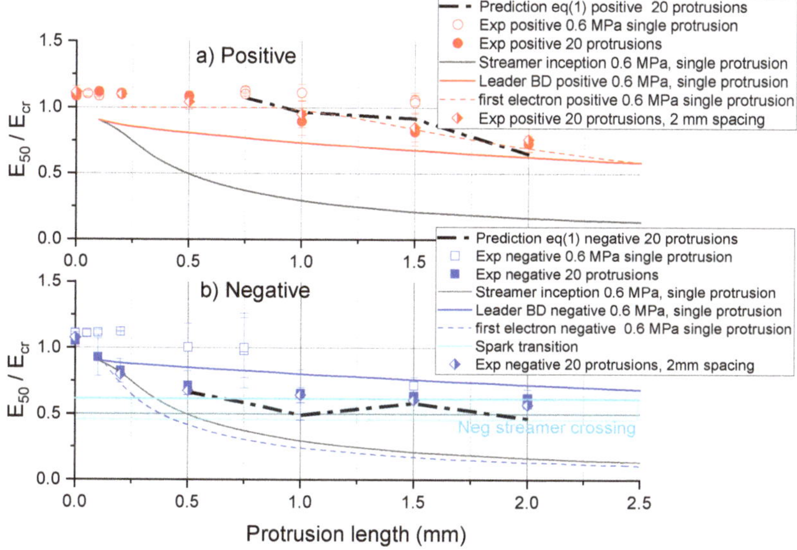

Figure 12. Normalized breakdown fields E_{50}/E_{cr} vs. protrusion length L for CO_2 and 20 protrusions at (**a**) positive and (**b**) negative polarities.

As mentioned in Section 2.1, also a smaller spacing of 2 mm between the protrusions was tested with 20 protrusions for CO_2. These results were not significantly different from the usual spacing of

4 mm—see Figure 12. Only a small reduction of less than 6% in field maximum was expected from the comparison of electric field calculations when using the smaller spacing (not shown in Figure 12), which is similar to the experimental uncertainties.

For 100 protrusions and positive polarity—see Figure 13a—the measured breakdown fields drop to the leader breakdown predictions for protrusions of 0.5 mm length and more. This is not predicted by the enlargement law, which results only in a moderate reduction. For smaller protrusions, the breakdown fields are still significantly higher than the leader breakdown prediction, probably due to the lack of a first electron. For larger protrusions of 1 mm and more, the positive breakdown fields are now similar or even lower than the negative ones, as would be expected from the leader breakdown criterion.

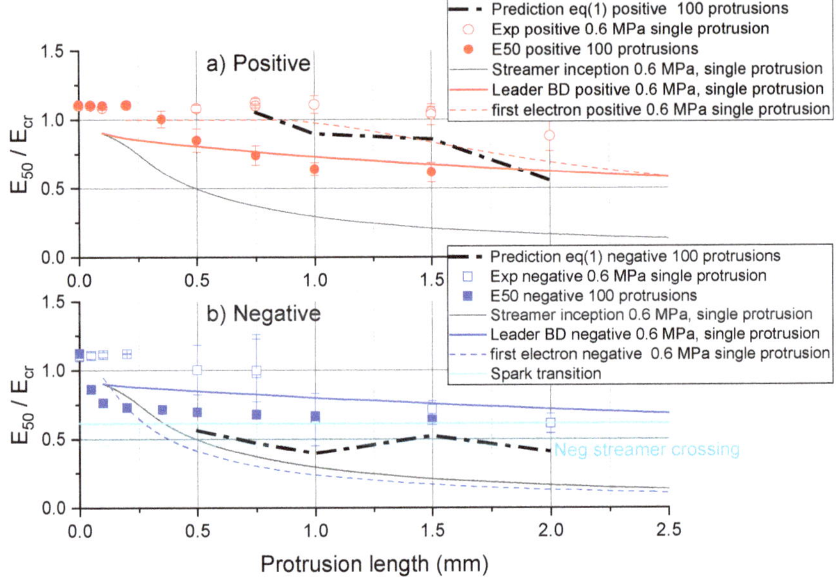

Figure 13. Normalized breakdown fields E_{50}/E_{cr} vs. protrusion length L for CO_2 and 100 protrusions at (**a**) positive and (**b**) negative polarities. Not all protrusion lengths could be measured when using multiple protrusions.

For 100 protrusions and negative polarity—see Figure 13b—the breakdown values drop significantly only for small protrusions below 0.5 mm lengths, compared to the previous case of 20 protrusions. For 250 μm and below, they are probably determined by streamer inception. Above 250 μm protrusion length, there is only a small drop of breakdown fields with increasing protrusion length. It seems that a saturation is reached, which is probably not due to the first electron criterion, but due to streamer crossing followed by spark transition. The enlargement law predictions are now significantly lower than the measurements.

Images of partial discharges and breakdown in CO_2 are shown in Figure 11. Local channel illuminations indicated local channel heating could be clearly observed, which might be during leader propagation but also for a streamer to spark transition. Similarly to SF_6, discharges occurred simultaneously in the time window of 10 μs. Weak density variations in the non-luminous regions indicate some channel heating also in these regions. Interestingly, breakdowns from the bottom plate electrode could also be observed.

4.3. Electrodes with Surface Roughness

The results from the electrodes with surface roughness are shown in Figures 14 and 15 for SF_6 and CO_2, respectively. Note that for the E_{50} evaluation the electric field at the surface of the electrodes was used, since the fields are weakly non-uniform in the case of the small surface. Again, the E_{50} breakdown field is normalized to the critical field. For comparison calculations for first electron, streamer inception and breakdown are shown for a single protrusion in the figures. Increasing the area in SF_6 at positive polarity led to a significant lowering of the E_{50} breakdown fields at positive, but negligible decrease at negative polarity—see Figure 14a. This can be understood by the streamer inception and first electron criterion. At positive polarity, the experimental breakdown field drops roughly to the predicted streamer inception field for large areas. This can probably be explained by a sufficiently high number of sites for discharge inception for large electrode areas, i.e., there is a high likelihood for a first electron at a small surface protrusion. Then, the streamer inception becomes a sufficient breakdown criterion. For small areas a first electron is probably lacking, and the breakdown field becomes even higher than the critical field. In the predictions for a single protrusion a first electron is only available at the critical field, i.e., probably these predictions are still too optimistic. The prediction from the enlargement law (solid arrow) for increasing area reproduces qualitatively the decrease of breakdown field with increasing surface area. However, it predicts for large areas the breakdown at the critical field and not at the streamer inception field. Using (1) to predict the breakdown field scaling from large to small area would quite underestimate the measured breakdown field (dashed arrow). Thus, at positive polarity, the enlargement law predictions do not agree well with the experiments.

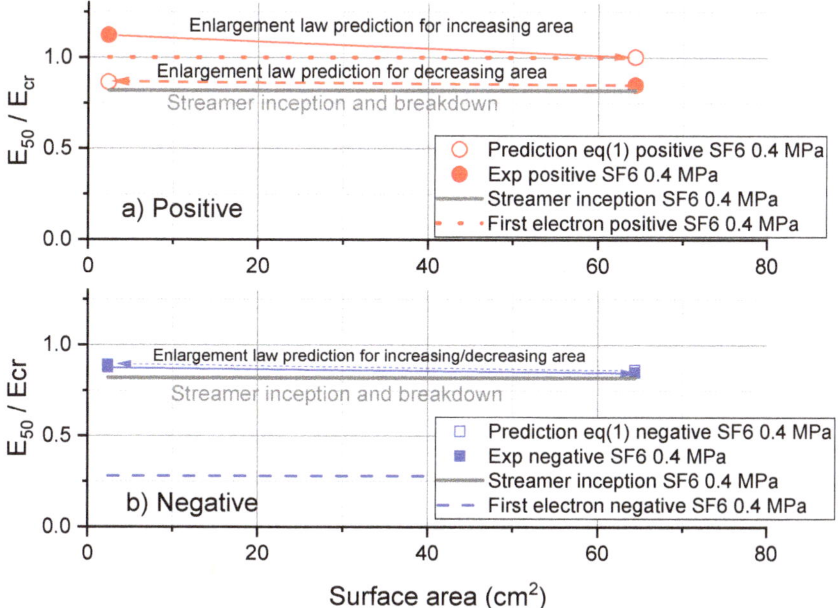

Figure 14. Normalized breakdown fields E_{50}/E_{cr} vs. effective surface area in SF_6 at 0.4 MPa at (**a**) positive and (**b**) negative polarities. Calculations for first electron, streamer inception and breakdown at a single hemispherical protrusion of 20 μm height are shown for comparison.

At negative polarity, this is different—see Figure 14b. Here, the predictions in both directions, i.e., from small to large electrode, or vice versa, give the correct breakdown fields. Note that the experimental breakdown field is in this case again approximately at the predicted streamer inception field. This is plausible, since at negative polarity there is no lack of a first electron (see dashed line in

Figure 14b) and breakdown occurs for small and large electrode roughly at similar fields, which is at streamer inception. Note that streamer inception is independent of polarity, i.e., for large areas, the breakdown fields are similar for both polarities.

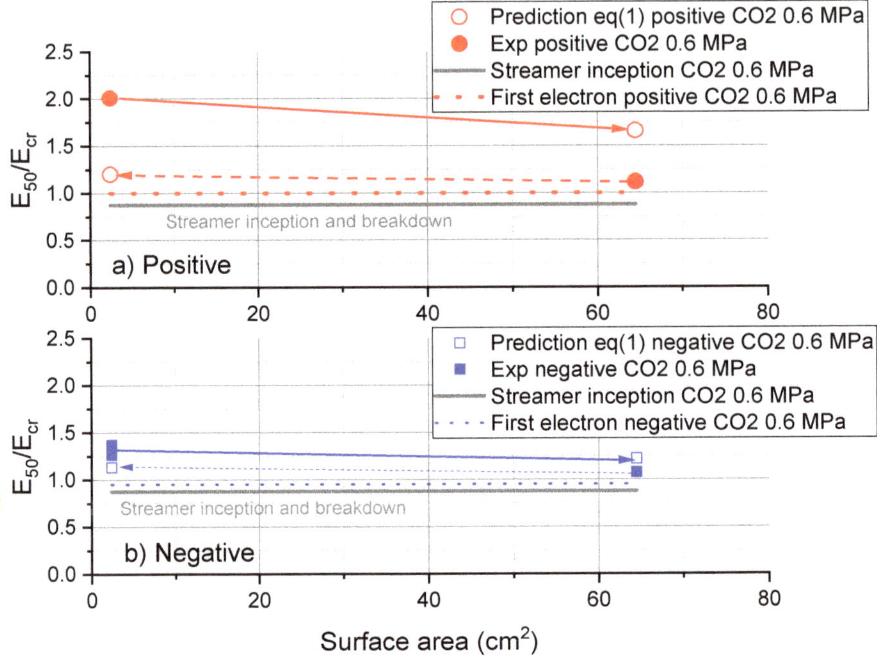

Figure 15. Normalized breakdown fields E_{50}/E_{cr} vs. effective surface area in CO_2 at 0.6 MPa at (**a**) positive and (**b**) negative polarities. Calculations for first electron, streamer inception and breakdown at a single hemispherical protrusion of 20 µm height are shown for comparison.

In CO_2, the observations are similar—see Figure 15. At positive polarity (Figure 15a), experimental breakdown fields drop for large electrodes to close to the critical field, which is slightly higher than the predicted streamer inception and breakdown field. This discrepancy is probably still within the accuracy of the predictions. Only for small area and positive polarity there is probably again a lack of a first electron which leads to an increase of E_{50} in the experiments. Note that this lack of first electron is not predicted by the model. Due to the strongly non-linear field dependence of (4), at the critical field, a first electron is always predicted. Probably the model is not sufficiently precise for CO_2 and small protrusions. At negative polarity in CO_2 (Figure 15b), there is a slightly more pronounced area effect than for SF_6. Again, the breakdown happens for large areas close to the critical field, which is only slightly above the first electron criterion prediction—see dashed horizontal line in Figure 15b. Thus, there is possibly also a lack of first electron for small areas at negative polarity, which is, however, much less pronounced than at positive polarity. The discrepancies can be probably explained by the simplicity of the model for the first electron.

Images of partial discharges and breakdown are shown in Figures 16 and 17 for both gases used. For CO_2, there are again discharges from the plate electrode, similarly to the protrusion results—see Figure 16. Arrested discharges could also be observed at breakdown for SF_6—see Figure 17. This shows that there is a competition between various discharge channels.

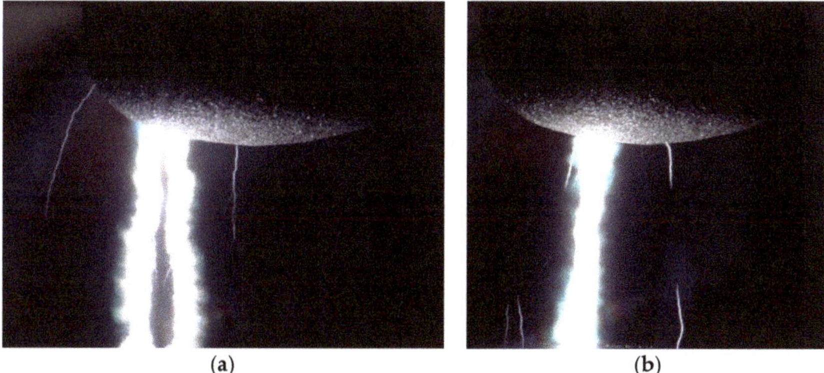

Figure 16. Images of partial discharges and breakdown in CO_2 at the small surface. (**a**) Positive polarity at a surface field of 322 kV/cm and (**b**) Negative polarity at 235 kV/cm surface field.

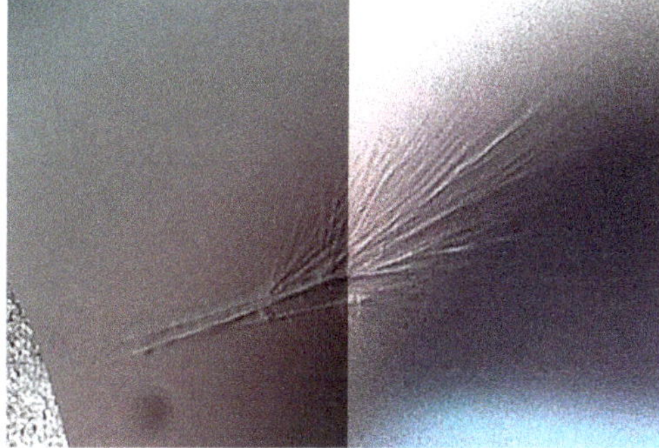

Figure 17. Images of partial discharges in SF_6 at 0.4 MPa, negative polarity at a surface field of 375 kV/cm. The picture is edited with different brightness on the two halves to better show the structures of the partial discharge.

5. Discussion

The results at multiple protrusions have shown that the experimental breakdown E_{50} values can be interpreted by the predictions of the various physical processes as first electron, streamer inception and leader propagation for a single protrusion. In SF_6, breakdown will only occur if all criteria are fulfilled. This is also valid in CO_2. However, error bars were much larger for the CO_2 experiments, resulting in larger uncertainties of the interpretations. In CO_2, streamer crossing and spark transition is a possible mechanism for breakdown at negative polarity, additionally to the leader breakdown, which is dominant at positive polarity. Similarly, this was found in [37]. In CO_2, breakdowns at protrusions occur much closer to the critical field, which is also reflected in the predictions. This is likely due to the lower slope of the effective ionization coefficient in CO_2, which is only a weakly electron attaching gas and, therefore, needs relatively higher fields for streamer and leader inception compared to the critical field than in SF_6 [50]. This leads to lower sensitivity to surface roughness or protrusions in case of CO_2 compared to SF_6.

Increasing the number of protrusions and comparing the resulting E_{50} breakdown fields with predictions from the enlargement law (1) leads to reasonable agreement for SF_6 at negative polarity. This indicates that the statistical processes for a single protrusion, resulting in the measured breakdown probability distribution, can be scaled according to the enlargement law without reaching other, e.g., physical limitations at the tested number of protrusions. This can be possibly understood by the very low first electron and streamer inception fields, i.e., these criteria are not limiting the scaling of the breakdown fields and the statistical scatter is probably mainly due to leader transition and propagation. This is different at positive polarity in SF_6. Whereas, the reduction of breakdown field with 20 protrusions is still well described by the enlargement law, we see a significant discrepancy using 100 protrusions. This is possibly produced by the availability of a first electron which needs a certain minimum field and can, therefore, not be described by statistical phenomena alone—see [34,38]. Thus, if the field is below a certain threshold using a large number of protrusions cannot compensate for the necessary field for creating the first electron in the critical volume. Note that the first electron depends on the size of the critical volume, which increases with the number of protrusions, but also strongly non-linearly on the applied electric field within the critical volume—see (4). Additionally, in CO_2, limits for the reduction of breakdown field with increasing number of protrusions can be seen. At negative polarity, such a limitation is the streamer inception criterion for small protrusions of 250 μm and smaller. For larger protrusions there is a lower limit which could be explained by streamer crossing and spark transition, but possibly also by the leader mechanism. This could not be determined unambiguously so far and would need more detailed optical diagnostic investigations. At positive polarity in CO_2, a reasonable agreement with the enlargement law is seen for 20 protrusions, which is similar to SF_6. For 100 protrusions and positive polarity the experimental breakdown fields drop to the predicted leader breakdown field. This drop is underpredicted by the enlargement law, which indicates that the breakdown probability distributions for single protrusions does not include all the relevant phenomena with sufficient precision in this case.

For rough surfaces, we investigated the case of very small protrusions in the range of only a few 10 μm in height. We used here the approach of Pedersen [14] to approximate surface roughness by hemispherical protrusions. Thus, this case has similarities to the protrusion array, only that the number of protrusions is probably much higher than in the case of the protrusion array. A detailed surface structural analysis was not done in the present case but unpublished investigations on a similar surface has shown that we can expect one protrusion of 20 μm height per mm^2 as an order of magnitude. For the small surface of 240 mm^2 we can expect, therefore, about 240 protrusions very roughly. For the larger area this was increased by more than a factor 20 to investigate the enlargement law in a more realistic situation. Thus, for the larger surface we can expect as an order of magnitude about 6450 protrusions of a 20-μm size. For negative polarity and SF_6, the area scaling due to this increase is satisfactorily described. This indicates that the probability distributions at negative polarity in SF_6 describe well the statistical phenomena. In this case, we do not expect a lack of a first electron and the statistical processes of the breakdown like streamer inception and leader transitions are probably decisive for the breakdown statistical scatter. In CO_2 and negative polarity, the reduction of breakdown field with increasing area is less well described by the enlargement law. Here, the statistical scatter of the first electron is also probably important. This is similar at positive polarity, where the breakdown fields in the experiments drop roughly to the streamer inception fields for large surfaces and the enlargement laws underpredict this. Possibly, statistical phenomena of the first electron do not follow a simple enlargement law scaling. This needs further investigations.

The investigation showed that enlargement laws cannot be applied without caution. The physical processes which might limit the scaling of statistical distributions, e.g., first electron or streamer inception, must be taken into account. Additionally, the enlargement law is sensitive to the precise shape of breakdown probability distributions, which might lead to insufficient accuracy if the distributions are not determined with sufficient precision. Finally, underlying physical processes might not follow a simple enlargement law scaling, e.g., due to strongly non-linear effects like for the first electron criterion. This is especially the case if various physical processes are involved.

In the present investigation, the lower limit of the Weibull distribution E_0 was set to the streamer inception values, which can be regarded as a minimum requirement for breakdown. As can be seen from the figures, in none of the cases this was reached with multiple protrusions. Only for rough surfaces and large area it was approached, i.e., in this case it influenced significantly the enlargement law. The satisfactory agreement for this case might indicate that this assumption for E_0 was justified.

The breakdown images have revealed some further information showing especially many discharges occurring in parallel during breakdown events on the short time scale of only a few microseconds. These discharges could bridge the gap or remain arrested in the gap. This shows that a first electron was not lacking in such cases, in agreement with the model predictions. Thus, at breakdown electrons were not only available at one protrusion but also at other sites. Interestingly, in CO_2 also discharges from the bottom plate could be sometimes observed at negative polarity, similar to a return stroke in lightning. This was not observed in SF_6. Possibly, this is linked to a different breakdown mechanism, which might be of streamer to spark transition type in this case.

6. Conclusions

Breakdown experiments on multiple protrusions and on artificially roughed surfaces were done in uniform and weakly uniform background fields with SF_6 and CO_2 at 0.4 and 0.6 MPa, respectively. The voltage waveform was a stepped DC pulse with time to breakdown within 10 μs, which is comparable to that of standard lightning impulse (LI). With the multiple protrusion array, the number of protrusions and the lengths of the protrusions was varied in the ranges of 1–100 and 0.05–2 mm, respectively. For the roughened surfaces, a mean peak-to-valley roughness (R_z) of 62 ... 65 μm was realized, and two different surface areas were tested. From the breakdown experiments, Weibull breakdown probability distributions were deduced and 50% breakdown probability fields and the standard deviation σ of the distributions were determined. Enlargement law scaling predictions according to (1) were done to predict the effect of changing the number of protrusions or surface area. These predictions were compared to the experiments. Additionally, calculations by physical models, describing first electron, streamer inception and breakdown by leader propagation or streamer-to-spark transition at single protrusions were done for interpretation of the experimental results. The interpretations were supported by optical observation of the discharges.

The following main observations were obtained: In SF_6, breakdown fields are well below the critical field, even for protrusions below 1 mm length, whereas for CO_2 breakdown occurs closer to the critical field. The breakdown fields can be interpreted and roughly described by the physical models, taking into account the uncertainties of the experiments and of the models. In SF_6 the leader propagation criterion is decisive for breakdown at both polarities for single protrusions. Increasing the number of the protrusions to 20 lowers the breakdown field for both gases in agreement with the predictions from the enlargement laws, compared to the single protrusion. No significant effect on the protrusions spacing was seen in CO_2 with 20 protrusions. Discrepancies with the enlargement law scaling were observed for both gases in some cases when increasing the number of protrusions to 100. The results show that enlargement laws cannot be generally applied without considering the various physical processes which influence breakdown, i.e., the availability of a first electron, streamer and leader inception and propagation and spark transition. The decisive breakdown criteria depend on the gas, the protrusion lengths and polarity of the applied voltage, which leads to different minimum limiting fields which cannot be exceeded by an enlargement law. Additionally, the various physical processes will not scale in the same way when using an enlargement law. All this limits the accuracy of an enlargement law scaling. This is confirmed by the results from the rough surfaces. At positive polarity the breakdown fields dropped for SF_6 and CO_2 approximately to the streamer inception field, when increasing the surface from 240 to 6450 mm^2, i.e., there is a physical limit for breakdown which is reached for sufficiently large surfaces. This decrease was underpredicted by the enlargement law. At negative polarity and SF_6, only a very small enlargement effect is seen in the experiments, in agreement with the predictions. In CO_2 and negative polarity, there is a more pronounced effect of the surface area, which is still much smaller than for positive polarity, however. Again, this decrease is underpredicted by the enlargement law. The effect of the protrusion spacing was investigated only in CO_2 with 20 protrusions. No significant effect within the experimental scatter was seen.

Author Contributions: Conceptualization, M.S. and O.C.F.; methodology, M.S., D.O. and O.C.F.; software, D.O. and O.C.F.; validation, M.S., F.M. and K.N.; formal analysis, O.C.F.; investigation, O.C.F.; resources, M.S. and D.O.; data curation, O.C.F.; writing—original draft preparation, M.S. and O.C.F.; writing—review and editing, M.S., O.C.F., K.N. and F.M.; visualization, O.C.F. and M.S.; supervision, M.S., F.M. and K.N.; project administration, O.C.F. All authors have read and agreed to the published version of the manuscript.

Funding: This research received no external funding.

Acknowledgments: The authors would like to thank Filip Halak for assisting with experimental work and Anders Ulfsnes for performing field calculations in COMSOL during their internships at Hitachi ABB Power Grids Research, Switzerland.

Conflicts of Interest: The authors declare no conflict of interest.

References

1. Ryan, H.M.; Jones, G.R. *SF_6 Switchgear*; Peter Pelegrinus Ltd.: London, UK, 1989.
2. Arora, R.; Mosch, W. *High Voltage and Electrical Insulation Engineering*; Wiley-IEEE Press: Piscataway, NJ, USA, 2011; ISBN 978-0-470-60961-3.
3. *IEC 62271-1 Ed. 1.0 2007-10 International Standard: High-Voltage Switchgear and Controlgear*; Part 1: Common Specifications; International Electrotechnical Commission: Geneva, Switzerland, 2007.
4. *IEEE C37.06-2009.09.11: AC High-Voltage Circuit Breakers Rated on a Symmetrical Current Basis-Preferred Ratings and Related Required Capabilities for Voltages above 1000 V*; IEEE: New York, NY, USA, 2009.
5. Mosch, W.; Hauschild, W. *Hochspannungsisolierungen Mit Schwefelhexafluorid*; VEB Verlag Technik: Berlin, Germany, 1979.
6. Kuechler, A. *High Voltage Engineering: Fundamentals-Technology-Applications*; Springer Vieweg: Heidelberg, Germany, 2017; ISBN 978-3-642-11992-7. [CrossRef]
7. United Nations Framework Convention on Climate Change. 2014. Available online: https://unfccc.int/ghg_data/items/3825.php (accessed on 4 April 2018).
8. Seeger, M.; Smeets, R.; Yan, J.; Ito, H.; Claessens, M.; Dullni, E.; Franck, C.M.; Gentils, F.; Hartmann, W.; Kieffel, Y.; et al. Recent development and interrupting performance with SF_6 alternative gases. *Electra* **2017**, *291*, 26–29.
9. Tian, S.; Zhang, X.; Cressault, Y.; Hu, J.; Wang, B.; Xiao, S.; Li, Y.; Kabbaj, N. Research status of replacement gases for in power industry. *AIP Adv.* **2020**, *10*, 050702. [CrossRef]
10. Li, X.; Zhao, H.; Murphy, A.B. SF_6-alternative gases for application in gas-insulated switchgear. *J. Phys. D Appl. Phys.* **2018**, *51*, 153001. [CrossRef]
11. Seeger, M. Perspectives on research on high voltage gas circuit breakers. *Plasma Chem. Plasma Process.* **2015**, *35*, 527–541. [CrossRef]
12. Rabie, M.; Franck, C.M. Assessment of Eco-friendly Gases for Electrical Insulation to Replace the Most Potent Industrial Greenhouse Gas SF_6. *Environ. Sci. Technol.* **2018**, *52*, 369–380. [CrossRef]
13. Nitta, T.; Yamada, N.; Fujiwara, Y. Area effect of electrical breakdown in compressed SF_6. *IEEE Trans. Power Appar. Syst.* **1974**, *2*, 623–629. [CrossRef]
14. Pedersen, A. Limitations of breakdown voltages in SF_6 caused by electrode surface roughness. In Proceedings of the Conference on Electrical Insulation & Dielectric Phenomena-Annual Report 1974, Downingtown, PA, USA, 21–23 October 1974; IEEE: New York, NY, USA, 1974; pp. 457–464.
15. Pedersen, A.; Karlsson, P.W.; Bregnsbo, E.; Nielsen, T.M. Anomalous breakdown in uniform field gaps in SF_6. *IEEE Trans. Power Appar. Syst.* **1974**, *6*, 1820–1826. [CrossRef]
16. Berger, S. Onset or breakdown voltage reduction by electrode surface roughness in air and SF_6. *IEEE Trans. Power Appar. Syst.* **1976**, *95*, 1073–1079. [CrossRef]
17. McAllister, I.W. A multiple protrusion model for surface roughness effects in compressed SF_6. *Elektrotechnische Z. A* **1978**, *99*, 283–284.
18. Crichton, B.H.; Lee, D.I.; Tedford, D.J. Prebreakdown in Compressed SF_6 and SF_6/N2 Mixtures in Projection Perturbed Uniform Fields. In Proceedings of the 4th International Conference on Gas Discharges, Swansea, UK, 7–10 September 1976; pp. 199–202.
19. Bortnik, I.; Ierusalimov, M.; Borin, V.; Varivodov, V.; Vertikov, V.; Ilenko, O.; Kuz'mina, Y. Evaluation of the requirements on the surface finish of electrodes in SF_6 insulation. *Electr. Technol. USSR* **1985**, *4*, 43–49.

20. Oiu, Y.; Li, R.; Kuffel, E.; Liu, M. Effect of electrode surface roughness on breakdown of SF_6 gas insulation. In Proceedings of the Conference Record of the 1988 IEEE International Symposium on Electrical Insulation, Cambridge, MA, USA, 5–8 June 1988; IEEE: New York, NY, USA, 1988; pp. 93–96. [CrossRef]
21. Li, R.; Qiu, Y. A new multi-ridge model for electrode surface roughness effect in SF_6. In Proceedings of the Second International Conference on Properties and Applications of Dielectric Materials, Beijing, China, 12–16 September 1988; pp. 105–108. [CrossRef]
22. El-Makkawy, S.M. Electrode surface roughness initiated breakdown in compressed SF_6 gas. In Proceedings of the Electrical Insulation and Dielectric Phenomena, IEEE 1994 Annual Report, Arlington, TX, USA, 23–26 October 1994; pp. 948–953. [CrossRef]
23. Lederle, C.; Kindersberger, J. The influence of surface roughness and coating on the impulse breakdown voltage in SF_6. In Proceedings of the 17th Annual Meeting of the IEEE Lasers and Electro-Optics Society, LEOS 2004, Boulder, CO, USA, 20 October 2004; pp. 522–525. [CrossRef]
24. Lederle, C.; Kindersberger, J. The influence of surface roughness and dielectric coating on ac and dc breakdown voltage in SF_6. In Proceedings of the 14th International Symposium on High Voltage Engineering, Beijing, China, 25–29 August 2005; pp. 1–6.
25. Shiiki, M.; Sato, M.; Hanai, M.; Suzuki, K. Dielectric performance of CO_2 gas compared with N_2 gas. In *Gaseous Dielectrics IX*; Christophorou, L.G., Olthoff, J.K., Eds.; Springer: Boston, MA, USA, 2001; pp. 365–370. [CrossRef]
26. Uchii, T.; Hoshina, Y.; Mori, T.; Kawano, H.; Nakamoto, T.; Mizoguchi, H. Investigations on SF_6-free gas circuit breaker adopting CO_2 gas as an alternative arc-quenching and insulating medium. In *Gaseous Dielectrics X*; Christophorou, L.G., Olthoff, J.K., Vassiliou, P., Eds.; Springer: Boston, MA, USA, 2004; pp. 205–210. [CrossRef]
27. Juhre, K.; Kynast, E. High pressure N_2, N_2/CO_2 and CO_2 gas insulation in comparison to SF_6 in GIS applications. In Proceedings of the XIVth International Symposium on High Voltage Engineering, Beijing, China, 25–28 August 2005.
28. Meijer, S.; Smit, J.J.; Girodet, A. Comparison of the breakdown strength of N_2, CO_2 and SF_6 using the extended up-and-down method. In Proceedings of the 2006 IEEE 8th International Conference on Properties applications of Dielectric Materials, Bali, Indonesia, 26–30 June 2006; pp. 653–656. [CrossRef]
29. Okabe, S.; Goshima, H.; Tanimura, A.; Tsuru, S.; Yaegashi, Y.; Fujie, E.; Okubo, H. Fundamental insulation characteristic of high-pressure CO_2 gas under actual equipment conditions. *IEEE Trans. Dielectr. Electr. Insul.* **2007**, *14*, 83–90. [CrossRef]
30. Hikita, M.; Ohtsuka, S.; Yokoyama, N.; Okabe, S.; Kaneko, S. Effect of electrode surface roughness and dielectric coating on breakdown characteristics of high pressure CO_2 and N_2 in a quasi-uniform electric field. *IEEE Trans. Dielectr. Electr. Insul.* **2008**, *15*, 243–250. [CrossRef]
31. Ka, S.; Inami, K.; Hama, H.; Ueta, G.; Wada, J.; Okabe, S. Lightning impulse and ac breakdown characteristics of CO_2 gas under quasi-uniform electric field using real-size gas-insulated switchgear model. In Proceedings of the 18th International Symposium on High Voltage Engineering, Seoul, Korea, 25 August 2013; pp. 1549–1553.
32. Yoshida, T.; Ka, S.; Shimizu, Y.; Inami, K.; Hama, H.; Ueta, G.; Wada, J.; Okabe, S. Metallic particle motion and its sparkover property at AC voltages in N_2, CO_2, dry air and SF_6. In Proceedings of the 9th International Workshop on High Voltage Engineering (IWHV 2014), Okinawa, Japan, 7–8 November 2014.
33. Yoshida, T.; Shimizu, Y.; Inami, K.; Hama, H.; Ueta, G.; Wada, J.; Okabe, S. Partial discharge and breakdown properties in N_2, CO_2, dry air and SF_6 initiated by metallic particles fixed on high voltage conductor and attached on spacer surface at ac voltages. In Proceedings of the 20th International Symposium on High Voltage Engineering, Buenos Aires, Argentina, 28 August–1 September 2017.
34. Bujotzek, M.; Seeger, M. Parameter dependence of gaseous insulation in SF_6. *IEEE Trans. Dielectr. Electr. Insul.* **2013**, *20*, 845–855. [CrossRef]
35. Seeger, M.; Niemeyer, L.; Bujotzek, M. Leader propagation in uniform background fields in SF_6. *J. Phys. D Appl. Phys.* **2009**, *42*, 185205. [CrossRef]
36. Seeger, M.; Stoller, P.; Garyfallos, A. Breakdown fields in synthetic air, CO_2, a CO_2/O_2 mixture, and CF4. *IEEE Trans. Dielectr. Electr. Insul.* **2017**, *24*, 1582–1591. [CrossRef]
37. Seeger, M.; Avaheden, J.; Pancheshnyi, S.; Votteler, T. Streamer parameters and breakdown in CO_2. *J. Phys. D Appl. Phys.* **2017**, *50*, 1–15. [CrossRef]

38. Seeger, M.; Niemeyer, L.; Bujotzek, M. Partial discharges and breakdown at protrusions in uniform background fields in SF_6. *J. Phys. D Appl. Phys.* **2008**, *41*, 185204. [CrossRef]
39. Hauschild, W.; Mosch, W. *Statistical Techniques for High-Voltage Engineering*; Peter Peregrinus on Behalf of the Institution of Electrical Engineers: London, UK, 1992; Volume 13, ISBN 978-0863412059.
40. Dunz, T.; Fruth, B.; Niemeyer, L.; Ullrich, L.; Diederich, K.; Hassig, M. Electrical field on rough electrode surfaces and its influence on the statistical properties of SF_6 breakdown. In Proceedings of the 6th International Symposium of High Voltage Engineering, ISH, New Orleans, LA, USA, 28 August–1 September 1989.
41. Weibull, W. A statistical distribution function of wide applicability. *Trans. Am. Soc. Mech. Eng.* **1951**, *32*, 293–297.
42. Kiyan, T.; Ihara, T.; Kameda, S.; Furusato, T.; Hara, M.; Akiyama, H. Weibull Statistical Analysis of Pulsed Breakdown Voltages in High-Pressure Carbon Dioxide Including Supercritical Phase. *IEEE Trans. Plasma Sci.* **2011**, *39*, 1729–1735. [CrossRef]
43. Whitehouse, D.J. *Surfaces and Their Measurement*; Kogan Page Science: London, UK, 2002; ISBN 978-1-903996-01-0. [CrossRef]
44. Turnbull, B.W. Nonparametric estimation of survivorship function with doubly censored data. *J. Am. Stat. Assoc.* **1974**, *69*, 169–173. [CrossRef]
45. Zaengl, W.S.; Petcharaks, K. Application of Streamer Breakdown Criterion for Inhomogeneous Fields in Dry Air and SF_6. In *Gaseous Dielectrics VII*; Christophoru, L.G., James, D.R., Eds.; Springer: Bosten, MA, USA, 1994; pp. 153–159.
46. Naidis, G.V. Simulation of streamer-to-spark transition in short non-uniform air gaps. *J. Phys. D Appl. Phys.* **1999**, *32*, 2649–2654. [CrossRef]
47. Seeger, M.; Votteler, T.; Pancheshnyi, S.; Carstensen, J.; Garyfallos, A.; Schwinne, M. Breakdown in CO_2/O_2 and $CO_2/O_2/C_2F_4$ mixtures at elevated temperatures in the range 1000–4000 K. *Plasma Phys. Technol.* **2019**, *6*, 39–42. [CrossRef]
48. Ono, R.; Oda, T. Visualization of Streamer Channels and Shock Waves Generated by Positive Pulsed Corona Discharge Using Laser Schlieren Method. *J. Appl. Phys.* **2004**, *43*, 321–327. [CrossRef]
49. Bujotzek, M.; Seeger, M.; Schmidt, F.; Koch, M.; Franck, C. Experimental investigation of streamer radius and length in SF_6. *J. Phys. D Appl. Phys.* **2015**, *48*, 245201. [CrossRef]
50. Seeger, M. Electric breakdown in high voltage gas circuit breakers. In Proceedings of the 22nd International Conference on Gas Discharges and Their application, Novi Sad, Serbia, 7 September 2018.

Article

Ablation-Dominated Arcs in CO_2 Atmosphere—Part I: Temperature Determination near Current Zero

Ralf Methling [1,*,†], Alireza Khakpour [1,†], Nicolas Götte [2,†] and Dirk Uhrlandt [1,†]

1 Leibniz Institute for Plasma Science and Technology (INP), Felix-Hausdorff-Str. 2, 17489 Greifswald, Germany; as.khakpour@gmail.com (A.K.); uhrlandt@inp-greifswald.de (D.U.)

2 Institute for High Voltage Technology, RWTH Aachen University, Schinkelstrasse 2, 52056 Aachen, Germany; goette@ifht.rwth-aachen.de

* Correspondence: methling@inp-greifswald.de; Tel.: +49-3834-554-3840

† These authors contributed equally to this work.

Received: 11 August 2020; Accepted: 9 September 2020; Published: 10 September 2020

Abstract: Wall-stabilized arcs dominated by nozzle–ablation are key elements of self-blast circuit breakers. In the present study, high-current arcs were investigated using a model circuit breaker (MCB) in CO_2 as a gas alternative to SF6 (gas sulfur hexafluoride) and in addition a long polytetrafluoroethylene nozzle under ambient conditions for stronger ablation. The assets of different methods for optical investigation were demonstrated, e.g., high-speed imaging with channel filters and optical emission spectroscopy. Particularly the phase near current zero (CZ) crossing was studied in two steps. In the first step using high-speed cameras, radial temperature profiles have been determined until 0.4 ms before CZ in the nozzle. Broad temperature profiles with a maximum of 9400 K have been obtained from analysis of fluorine lines. In the second step, the spectroscopic sensitivity was increased using an intensified CCD camera, allowing single-shot measurements until few microseconds before CZ in the MCB. Ionic carbon and atomic oxygen emission were analyzed using absolute intensities and normal maximum. The arc was constricted and the maximum temperature decreased from >18,000 K at 0.3 ms to about 11,000 K at 0.010 ms before CZ. The arc plasma needs about 0.5–1.0 ms after both the ignition phase and the current zero crossing to be completely dominated by the ablated wall material.

Keywords: circuit breaker; switching arc; optical emission spectroscopy; ablation; current zero; SF6 alternative gases; CO_2; PTFE

1. Introduction

For modern power transmission and distribution grids, high voltage circuit breakers are among the essential elements to ensure safe power flow [1,2]. Basic technology applied therefore are self-blast circuit breakers in which a pressure build-up in a heating volume, necessary for arc quenching around current zero (CZ), is produced by the ablation of material from the nozzle wall due to intense arc radiation. Usually, gas sulfur hexafluoride (SF_6) is applied as quenching and insulating gas due to its unique properties as being chemically inert, non-flammable, non-explosive, non-toxic, thermally stable, and an excellent electrical insulator and arc interrupter due to its high electronegativity (electron attachment) and density [3]. Metal-doped polytetrafluoroethylene (PTFE) is used as the nozzle material due to well-adjustable ablation, pressure built-up, and dielectric properties. A main trend of circuit breaker development is the substitution of the extremely potent greenhouse gas SF_6 with a high global warming potential (of about 23,000 times that of CO_2 over a 100 year period) by more environmentally-friendly gases [4]. A variety of alternative gases has been discussed and tested in the last decades, e.g., CO_2, CF_3I, C_2F_4, c-C_4F_8, C_4F_7N, and C_5F_{10} as pure gases or in mixtures of two or three gases including components like N_2, O_2, and CO_2. However, only a limited selection remains in the actual investigations. Gas mixtures containing fluoro–nitriles (C_4F_7N) or fluoro–ketones ($C_5F_{10}O$) as minority components (<20%) have been identified as the most promising alternatives to SF_6 in high-voltage gas-insulated switchgear applications because of their low global warming potential together with their dielectric strength values being comparable to SF_6 [5–7]. Due to high boiling points (at 1 bar) of the fluoro–ketones (27 °C) and fluoro–nitriles (−4.7 °C), gas mixtures with carbon dioxide CO_2 as buffer gas prevent liquefaction at temperatures below −30 °C. One of the most promising alternatives that has less limitations concerning temperature range and greenhouse effect but reasonable electrical insulation is CO_2 [8–10]. Moreover, some of the experimental techniques and results obtained for CO_2 will be relevant for other alternative gases that are usually applied with CO_2 as buffer gas.

For an evaluation of the interruption performance it is mandatory to analyze and understand the time around CZ, i.e., the phase of current interruption and recovery of dielectric insulation in the electrode gap. Beside experimental investigations, computer simulations are of a high importance due to their cost-efficiency and fast adaptation to different geometries. However, such simulation tools need to be validated with experimental results and also to be provided with reliable input parameters based on experimental data. From the physical point of view, numerous transient effects can be observed such as:

- Reversal of the gas flow in the heating channel,
- Transition from an ablation-controlled to an axially blown arc,
- Arc constriction and finally extinction,
- Wall ablation that still occurs when the energy input by radiation from the arc is terminated, and
- Dielectric recovery of the electrode gap region.

Hence, the experimental investigation of the CZ phase is of high relevance for the development of high voltage switchgear. One of the main goals is the determination of physical properties as the composition, pressure, and temperature of the plasma. Ideally, this should be done with both spatial and temporal resolution as close as possible to CZ. Such investigations are usually based on optical methods and demand view ports or other access to the arc plasma that is often not available for commercial circuit breakers. A balance must be found between a conservation of geometry, functionality, and plasma conditions on the one side and a modification of the setup on the other side allowing, e.g., optical access as well as fast exchange of fill gas and components as electrodes, nozzles, windows, and ignition wires. In general, model circuit breakers are often equipped with fixed electrodes and ignited by an explosion of thin wires [11]. Experiments with moving electrodes of a pin and tulip shape may provide higher similarity with commercial circuit breaker geometries including the geometry of nozzles, heating chambers, and gas flow. However, reproducibility is often critical for experiments with moving electrodes. Furthermore, often the optical access is either fairly sophisticated and limited in time (using slits and windows [12]) or limited in spatial resolution, e.g., using optical fibers. If side-on diagnostics of the arc through quartz windows placed outside the nozzle is not possible due to non-transparent vapor, an alternative end-on arc observation can be realized through a ring electrode [13]. Alternatively, the PTFE as the standard nozzle material in commercial circuit breakers can be replaced by polymers that are transparent, e.g., PMMA [14,15].

In this work, two setups are used for investigation with optical emission spectroscopy (OES). In both setups, two pin electrodes are placed in a fixed distance and surrounded by PTFE nozzles. Ignition wires are used to initiate the arcs. The main difference between the setups is the nozzle geometry:

- The first, more simple setup is applied to generate extra-high pressure built-up and strong influence of the wall ablation. Therefore, the nozzle is made of a single, long PTFE tube. The influence of ignition wire and surrounding gas (ambient air) during the early stage of the arc discharge have been investigated in [16]. In the present work, the focus is set on the gas flow reversal as well as the detection limits for the determination of plasma temperature profiles around CZ. For these issues, no surrounding chamber is needed.

- The second setup is a model self-blast circuit breaker in a CO_2 atmosphere with optical access via the windows. It consists of two nozzles surrounding the electrodes and forming a heating channel which leads into the heating volume. Earlier experiments were carried out with the optical observation at the position of the heating channel and in the high-current phase [17]. Hence, the plasma emission from the central parts of the arc is influenced by the axial flow of hot gas in the heating channel, i.e., along the line of sight. The gas flow into the heating volume partly forces the plasma into the heating channel and the emission region exceeds the nozzle diameter, showing turbulence and deviation from the expected bi-convex structure that is needed for plasma temperature determination. To overcome these problems, for the experiments reported here the observation position was shifted away from the heating channel and towards the electrodes. As described below, this was realized by insertion of quartz windows into the nozzles.

For the investigation of the arc plasma, OES of atomic and ionic lines is a standard method for experimental determination of temporally- and spatially-resolved profiles of composition, temperature, and partial pressures that are needed for the calculation of thermal and electrical conductivity. However, the applicability of optical emission spectroscopy is limited for low emission levels, e.g., at low currents around CZ or at positions near the nozzle wall. Usually, plasma temperatures below 8000 K cannot be measured due to low line intensity and limited sensitivity of the detector. Thus, spatial temperature profiles cannot be obtained in the close vicinity of the wall region that is of high interest due to the importance of the nozzle ablation [18,19].

An alternative to OES at lower temperatures could be optical absorption spectroscopy (OAS) of resonant lines since the majority of atoms is in the ground state and thus, detectable by absorption. However, corresponding investigations under switching-relevant conditions are rather sophisticated; most resonant lines of carbon, fluorine, and copper need an optical access in the ultraviolet (UV) wavelength region whereas others like O I at 630.03 nm are very weak. Alternatively to atomic and ionic lines, molecules might be investigated. Most molecules dissociate above critical temperatures below 10 000 K. Thus, such analysis has the potential to provide insight into the region close to the wall at lower temperature. Experiments on OAS and investigation of molecule emission and absorption are reported in an accompanying article [20].

In the present paper, several questions should be answered: What can be learned by application of different experimental techniques for instigation of ablation-dominated arcs, partly combined? How far to current zero plasma temperature profiles can be measured with OES? What happens in the low-current phase approaching CZ with these profiles when lower conductivities are needed—mainly a temperature decrease or constriction of the arc diameter?

2. Materials and Methods

2.1. Geometry of Electrodes and Nozzles

As described above, two setups of electrodes and nozzles are used. They are sketched in Figure 1. In both cases, two-pin tungsten-copper (W-Cu) electrodes of 10 mm diameter were placed horizontally with a fixed contact distance of 40 mm. The electrodes were surrounded by either one 126 mm long, tubular-shaped nozzle of 50 mm outer diameter (a), or two separate nozzles of about a 50 mm length and 104 mm outer diameter (b). The general structure of the tubular and separated nozzles were similar. They were made of PTFE doped with <0.5 wt% molybdenum disulfide (MoS_2) as usually used in high-voltage circuit breakers (CB). In both cases, the inner diameter was about 12 mm in the central part where the arc discharge was burned and about 16 mm at the electrode side ends of the nozzle to allow an exhaust gas flow along the electrodes. In the vicinity of the electrode tips, a smooth transition was realized from the smaller to the larger diameter.

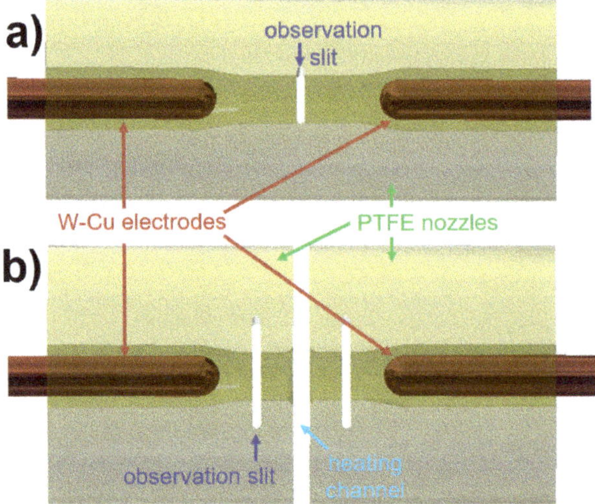

Figure 1. Setups (**a**) with a closed, long PTFE nozzle for experiments with strong ablation and high pressure built-up and vertical observation slits in the middle and (**b**) with two separated PTFE nozzles forming a heating channel for plasma flow into a heating volume as used for the model circuit breaker.

For the camera observation and spectroscopy, optical access was realized by vertical slits of a 2 mm width. Pairwise placement at opposite positions allowed not only an observation of the emitted radiation but also background illumination and absorption measurement. The slits were mortised directly into the nozzles, ranging over the complete nozzle diameter. Since such slits would be potential exhaust pipes of the plasma causing additional disturbances, 2 mm–thick quartz plates were applied to seal the nozzle. After each shot the sealing plates were checked visually and the transmission was measured regularly. Because of the possible fume deposition on the plates, they were exchanged after each shot but could be recycled after cleaning. With regular exchange, no melting of quartz glass was observed. Generally, it was found that the sealing was very effective and the transmission reduction was controllable. However, few cases were observed with severe, local blackening of the quartz plates indicating problems with the sealing and these shots were repeated. Breaking of the plates occurred only for ones after extremely high currents.

For setup (a) the observation slits were placed in the middle between both electrodes, i.e., about 20 mm away from both electrodes. In setup (b), the distance between the two nozzles was about 4 mm forming a heating channel. The line of sight used in [17] was along the heating channel, i.e., in a central position between the nozzles where turbulent gas flow in a radial direction cannot be avoided and may disturb the observation of radial profiles. Having the opportunity to use quartz-sealed vertical slits, the observation point was positioned in one of the nozzles (the nozzle on the left side) in Figure 1. It was at half distance to the electrode tip and the nozzle exhaust in the heating channel, i.e., about 9 mm away from both.

The arcs were operated under ambient conditions in case of setup (a), i.e., without external chambers. The apparatus shown in the left part of Figure 2 basically comprises of electrically-isolated holders for nozzle and electrodes including ceramic shielding protecting against the hot exhaust plasma. Setup (b) was part of a model circuit breaker that is shown in Figure 2. It was placed in an outer vessel of about 300 L volume that was evacuated and filled with CO_2 before each shot. Additional windows in the model chamber and the vessel allowed a free view through the nozzle. Thus, in this case an observation was also possible from opposite directions as well as absorption measurements.

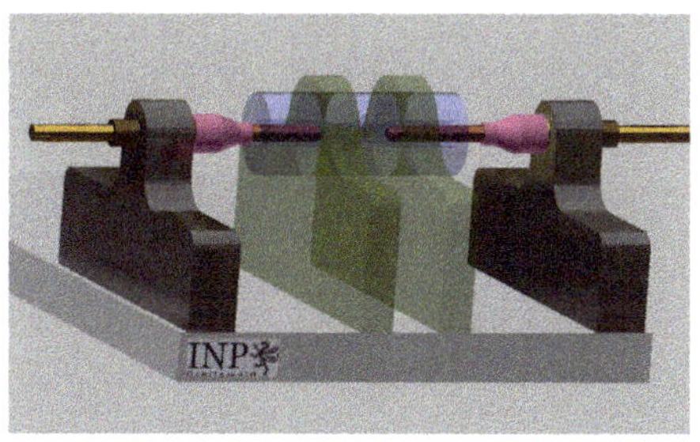

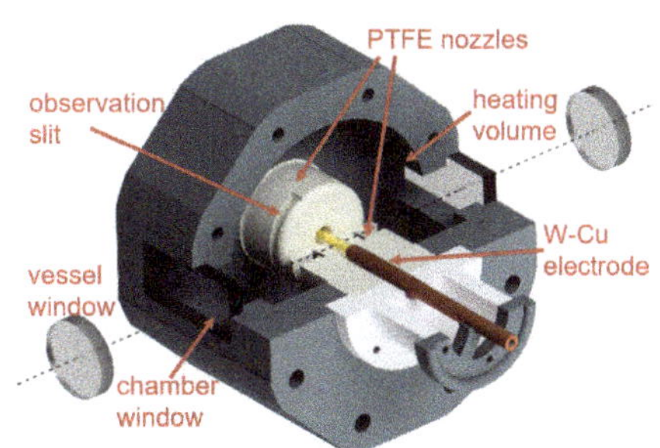

Figure 2. Left: Setup (a) was used under ambient conditions (no external chamber). **Right**: Setup (b) was part of a model circuit breaker placed inside a pressure vessel with vessel windows and the quartz plates in the observation slits. The dotted line indicates the sight through the outer vessel.

LC circuits were used to generate sine-like current waveforms of about a 100 Hz frequency and 11 kA peak current for setup (a) and 50 Hz and 5.3 kA for setup (b). The arc discharges were initiated using exploding thin Cu wires. Pressure sensors (603A from Kistler) were positioned in case of setup (a) in the middle of the nozzle, i.e., in 90° to the observation slits. In case of setup (b) the pressure sensor was placed in the heating volume of the model circuit breaker.

2.2. Optical Setup

Different methods were applied for the optical analysis. Firstly, high-speed cameras (HSC), the 24bits color Y6 or the 10bits monochrome Y4, both from Integrated Design Tools (IDT), were used to observe the general discharge behavior. Although the available area was reduced to a 2 mm-wide slit, relevant information was obtained regarding gas flow, asymmetry, dust, and droplets. As an example, a special double frame optics (DFO) could be applied between camera and lens that divided the optical path into two ways and produced two images of the same area that could be filtered at different wavelengths.

Secondly, optical emission spectroscopy was carried out by means of an imaging spectrograph with a 0.5 m focal length (Roper Acton SpectraPro SP2500i). The nozzle slit was imaged on the entrance slit of the spectrograph by a focusing mirror to spectrally investigate arc cross sections, i.e., perpendicular to the arc axis. Having the development during arc discharge in the focus, the spectrograph was equipped with a high-speed video camera (Y4 series of IDT). That enabled to record a series of 2D-spectra with typical repetition rates of 100 µs, allowing approximate exposure times of 98 µs (frame rate of 10,000 fps). It should be noted that rather long exposure times were necessary due to limited sensitivity of non-intensified HSC. Alternatively, the HSC was replaced by an ICCD camera (PI-MAX4 from Princeton Instruments) with higher sensitivity, allowing only single images but of shorter exposure times and at lower intensities, e.g., around current zero.

In a compromise between light intensity, spectral resolution, and exposure time, the entrance slit of the spectrograph was set to 50 µm. With gratings of 150 lines per mm for overview and 1800 L/mm for detailed spectra, the spectral range was 150 nm and 10 nm, and a spectral resolution (full width half maximum) of 0.7 nm and <0.1 nm, respectively. The intensity of side-on spectra was calibrated in units of spectral radiance by means of a tungsten strip lamp (OSRAM Wi 17/G) at the arc position. The window transmission of 50–70% was taken into account, mainly resulting from the coating of the quartz plates at the nozzles.

A positioning laser was used to adjust line of sight and check the stability between the shots. However, it was found that the width of 2 mm of the observation slit at the nozzle allowed a stable and free optical path. Usually, no corrections between the shots were needed thanks to the absence of moving parts.

3. Results

3.1. Video Observation and Flow Reversal

On the left side of Figure 3, a typical current waveform (red) is shown for the discharges in the model circuit breaker with a 5.3 kA peak value (setup (b)). The current does not have the exact shape of a perfect sine and has a slightly longer duration than expected for 50 Hz (∼10.6 ms). In this article the main focus is set on the time around CZ while the starting phase is of minor interest. Consequently, all points in time are given with respect to the current zero crossing for easier comparison. The voltage curve (blue) is characterized by a long period of values around 200 V flanked by two distinct peaks at the explosion of the ignition wire and at the arc extinction near CZ. During the high-current phase of arc discharge, the voltage increases from the local minimum of about 140 V at the end of the ignition peak (10 ms) to a maximum of about 240 V during the peak current (5 ms), and slightly decreases again with lower currents until about 200 V (3 ms before CZ).

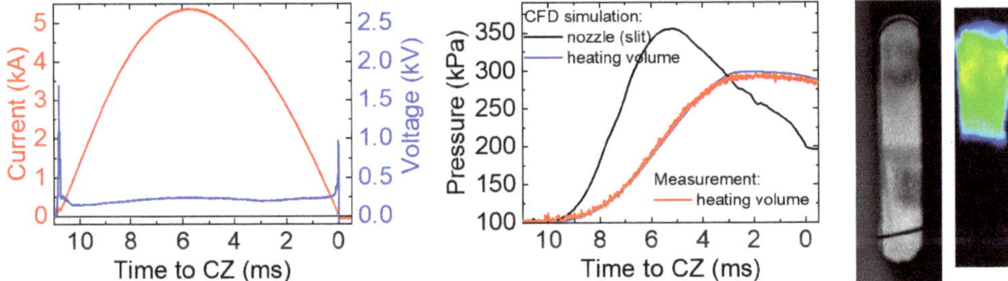

Figure 3. Left: Current and voltage waveform—nearly a sine 50 Hz shape and duration (∼10.6 ms). **Middle**: Pressure simulation for two positions and measurement for heating volume. **Right**: Photo (grey) of window with ignition wire and example for observation of the general behavior of the discharge about 1 ms before current zero (CZ) (false color representation) showing an asymmetric position of a constricted arc.

The intense radiation emitted from the arc plasma is absorbed by the surface of a surrounding PTFE nozzle where it causes photo–ablation of the wall material [13]. Part of the ablated material is exhausted as vapor by the axial flow, the other part enters the arc and is heated to a plasma temperature by absorption of radiation coming from the arc interior, leading to a discharge that is dominated by the ablation [18]. Both the arc plasma and vapor create an overpressure which causes an axial expansion flow towards the ends of the nozzle. In the middle of Figure 3, three curves of the pressure are shown. The red and black curve were obtained for the slit position inside the PTFE nozzle and the heating volume, respectively, from CFD simulation (CFD-ACE+ software suite by the ESIGroup) of the discharge according to [17]. A comparison of simulation results for the heating volume (black) with our measurements (dashed blue) resulted in a good agreement concerning the shape of the curve. Small deviations of the absolute values were used to adjust the pressure in the model. Although no pressure sensors could be placed directly in the nozzle or heating channel in case of the MCB, values of the pressure at the OES position (slit) could be easily obtained from a comparison with the results from the CFD simulation (red curve). The maximum pressure of 3.5 times the filling pressure is built up close to the peak current. It should be noted that at CZ the pressure is still about two times the filling pressure, which results in a cooling of the arc.

On the right part of Figure 3, a grey scale image is shown to illustrate the area of HSC observation at the position of the nozzle slit. Note that illumination from the backside was applied and the ignition wire can be recognized in the lower part. Finally, an example of a HSC frame of the arc in false colors is given on the far right about 1 ms before CZ. Typically, the arc is constricted at this time to less than half of the nozzle diameter. Here, it is located in the upper part of the nozzle, showing a deviation from rotational symmetry.

The arc in the slit area was further investigated using the double frame optics (DFO). Filters were applied to reveal information on the plasma composition. Any emission from fluorine and oxygen can be clearly dedicated to the nozzle wall (PTFE, C_2F_4) and filling gas (CO_2), respectively. Therefore, the left channel was equipped with a metal interference filter (MIF) with maximum transmission at 675 nm to become sensitive for several atomic fluorine lines F I and the right channel with a narrow filter at 777 nm for the atomic oxygen triplet O I. The emission from carbon lines could not be used since both materials contained carbon. On the upper part of Figure 4, six exemplary double frames are shown for significant points in time. Please note that the full video is available as supplementary material to this article. In the lower part of Figure 4 the plasma composition as obtained from the CFD simulation is shown as fractions of PTFE and CO_2.

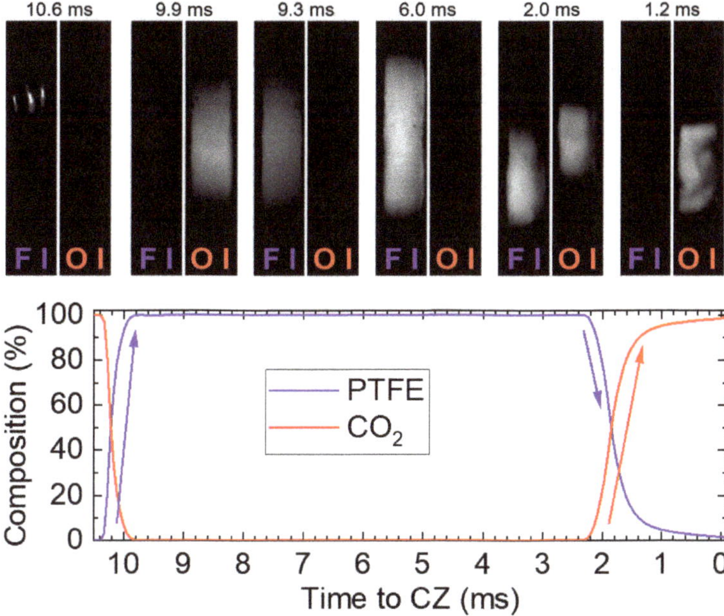

Figure 4. Visualization of the flow reversal before CZ for setup (b). **Top**: Selected frames of high-speed cameras (HSC) imaging using double frame optics; filters were applied for the atomic line emissions of F I 675 nm (left channel) and of O I 777 nm (right channel) to visualize emission from PTFE and CO_2, respectively. **Bottom**: Plasma composition as obtained from the CFD simulation.

The first frame from the DFO mainly shows the atomic copper emission from ignition wire that is present on the left side (Cu I line at 674.1 nm). The wire was divided by the explosion into three pieces within the 2 mm slit area. Less than a millisecond later, the emission dominated in the right channel, indicating a discharge in the CO_2 atmosphere (9.9 ms to CZ). Within few hundreds of microseconds, the brightness of the right channel faded while the left part became more intense. Here, the increasing wall ablation started to dominate the discharge, blowing the fill gas out of the nozzle. This is in good agreement with the simulation results. In the following, a long and stable period was observed that was dominated by ablation (cf. frame at 6.0 ms before CZ). Another reversal of flow was found about 2 ms before CZ: When the arc current and thus wall ablation were considerably decreased, the pressure in the nozzle also decreased to values below that of the heating volume, see Figure 3. As a consequence (relatively cold) gas from the heating volume with a high fraction of CO_2 flowed back into the nozzle. There it was heated up by the arc current and its radiation could be seen in the right channel. Interestingly, different intensity distributions were found for the left and right channel at 2.0 ms before CZ. That indicated non-homogeneous gas mixture. Finally, only emission from O I could be seen in the last ms, indicating a plasma composition completely dominated by CO_2.

3.2. OES of High-Current Phase Using High-Speed Camera

Series of two-dimensional spectra were obtained by means of video spectroscopy, allowing to investigate the different phases of the discharge within one acquisition (setup (b) with same conditions as above). This is illustrated in Figure 5 where six selected frames of one shot are shown exemplarily. The vertical dimension of each frame is related to the spatial distribution of the arc cross section along the nozzle slit whereas the spectral distribution is represented by the horizontal dimension. Please note that the full video is available as supplementary material to this article. The major lines and the time in relation to current zero are labeled above and on the left, respectively. For line classification, energy levels, transition probabilities, and other information on spectral lines the databases from NIST [21] and Kurucz [22] were applied. The spectra are dominated by atomic and ionic line emission from Cu, C, O, and F; the highest intensities are found in the central part of the axis. Temperature profiles can be determined along most of the arc radius based on these lines. In cases of significant nozzle ablation, usually F I or C II lines were utilized [11,12,17].

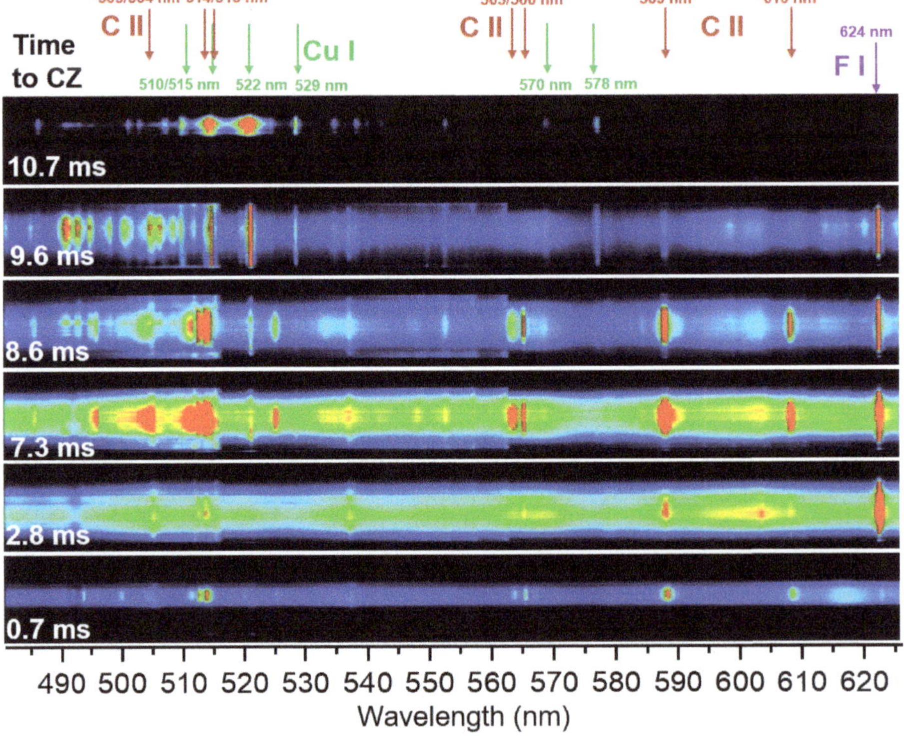

Figure 5. 2D-optical emission spectroscopy (OES) frames acquired at selected time points during one discharge by video spectroscopy with setup (b). From top to bottom: Ignition phase dominated by atomic Cu I lines from wire, transition to ablation-dominated arc with ionic C II and atomic F I lines from PTFE and close to CZ only C II from CO_2.

The first spectrum with detectable light emission acquired during the initial phase (10.7 ms before CZ) is characterized by the exploding ignition wire made of copper. Thus, only atomic Cu I lines are visible (labeled by green arrows). Obviously, the wire position was slightly off-axis for this shot. Although an influence of the ignition wire can be proved to last up to 1 ms, the bright peak is much shorter, typically below 100 µs [16]. Next, the filling gas CO_2 causes an additional emission of carbon ionic C II lines (red) and of the oxygen lines (O I at 777 nm, not shown here). This emission was found to be visible in OES after 100–220 µs. Under extreme conditions, i.e., for the setup (a) with the tubular PTFE nozzle exposed to high peak currents of 11 kA at 100 Hz, the lines of Cu I, C II, and O I can have intensity peaks already at 300–400 µs after ignition, while the F I emission (from PTFE) starts at about 300–400 µs after ignition [16].

Further emission of Cu I lines can be observed due to erosion of the W–Cu electrode, cf. the second spectrum at 9.6 ms before CZ. Here, the Cu I emission is still intense and distributed over the whole nozzle diameter. Additionally, rather weak carbon ionic as well as much brighter atomic fluorine lines (F I at 624 nm only in this spectral range) are visible. Although the influence of CO_2 is strongly decreased at 9.6 ms to CZ as it is known from the DFO investigation and plasma composition calculation, a ratio in particle density of about 1:2 of carbon to fluorine can be expected (C_2F_4 are the building blocks of PTFE). The transition probabilities of the line emission are roughly comparable, e.g., summarizing to about $10 \times 10^7\,s^{-1}$ for three C II lines at 609/610 nm and about $3 \times 10^7\,s^{-1}$ for the F I line at 624.0 nm. Considering the much stronger F I emission this means that excitation of F I with upper level at $\sim$14 eV is much more pronounced than an ionization of carbon atoms and excitation of C II with $\sim$24 eV.

The third spectrum at 8.6 ms before CZ is acquired in the phase of increasing PTFE nozzle ablation that causes a stronger emission of both atomic fluorine and carbon ionic lines. Furthermore, the C II and F I lines have comparable intensities, indicating more ionization and excitation of the C II levels than before and thus, considerable higher temperatures of the arc. The vanishing Cu I indicates that the metal vapor from electrode erosion is blown "backwards" by the new established flow of hot, ablated nozzle material.

Approaching the peak current (7.3 ms before CZ), continuum radiation can also be observed. On the one hand, there is a very broad continuum over the whole spectral range that is mainly emitting in the center of the arc. On the other hand, there is a pattern of many lines (often not resolved with this spectral resolution) of increasing intensities with increasing wavelengths until rather sharp edges, e.g., at 516 and 563 nm. This structure could be attributed to an emission of diatomic carbon molecules, i.e., the Swan band system arising from transitions between the $d^3\,\Pi_g$ and the $a^3\,\Pi_u$ electronic states of C_2 molecules. These Swan bands will be discussed later in detail. It should be noted that its spatial distribution ranges over the whole nozzle diameter, partially with peaks near to the nozzle walls.

With decreasing currents, the intensity of ionic C II lines decreases again, the atomic F I line at 624 nm has a much higher intensity at 2.8 ms before CZ whereas the C II lines are much weaker. Similar to the third spectrum at 8.6 ms to CZ, the excitation of ionic states is much less pronounced than that of fluorine atoms with upper levels around 14 eV due to a significantly lowered plasma temperature.

In the following phase of low current, the radiation emitted by the arc plasma no longer causes sufficient wall ablation to sustain the high pressure in the nozzle (cf. spectrum at 0.7 ms before CZ). The consequences are a reversal of the gas flow direction, disappearance of the fluorine line, and a spectrum dominated by C II line emission originating from the back-flowing CO_2 from the heating channel (plus continuum). Furthermore it should be noted that the arc is constricted, i.e., only a rather small part of the radial profile carries a contribution to emission and thus, also to electrical conductance.

3.3. *Optical Emission Spectroscopy near to Current Zero*

In the following the phase around current zero will be investigated for both nozzle geometries. Besides, the potential and limits of optical emission spectroscopy shall be discussed. The long, tubular nozzle of setup (a) is used to produce intense and short discharges (11 kA at 100 Hz) that will be analyzed by OES with a high-speed camera (HSC). Investigations with a more realistic setup (b) in the model circuit chamber (5 kA at 50 Hz) will be carried out using OES with an intensified CCD (ICCD) camera to come as close as possible to CZ.

3.3.1. Oes with High–Speed Camera

At first, the case of a current zero phase after an intense arc is considered. A strong PTFE wall ablation and pressure built-up is realized with a higher current of 11 kA peak. Using setup (a), the material flow is directed towards the exhausts at both electrode sides since the tubular nozzle is not interrupted, e.g., by a heating channel. Thus, the plasma is expected to remain relatively hot and dense for a longer time near CZ, at least in the observation position in the middle of the tube. It should be noted that the setup does not perform switching, i.e., the current is continued after CZ with reversed polarity. Additionally, there is only a short period of low current, i.e., it takes only $\sim$2.5 ms from peak to CZ for 100 Hz. In other words, the best conditions have been chosen for the analysis of the fluorine lines close to CZ.

Concerning imaging spectroscopy with high-speed camera, both temporal and spatial resolution are of interest. However, a limit was given by the detector's readout rate. Hence, the highest temporal resolution was obtained by a drastic reduction of the number of vertical lines, practically giving up the spatial dimension (side-on information). In the following, only the central position is used to analyze the temporal evolution of the spectrum with a frame rate of 100 kfps (10 µs repetition rate). In the upper part of Figure 6, the spectral evolution around CZ is shown. It should be mentioned that neither the full spectral range nor all spectra acquired during the discharge are plotted in order to focus on distinguishable lines of the relevant species during the phase around current zero. Some major lines of atomic fluorine as well as oxygen and nitrogen are labeled. Although there were no lines from carbon observed in the diagrammed spectral range (738–783 nm), other lines were checked and confirmed the finding described below, e.g., the ionic carbon lines C II 658 nm and 723 nm.

Among the 30 successive spectra shown in Figure 6, 17 were taken before and 13 were taken after CZ (see timescale at Y-axis and arrow). As expected for an ablation-dominated arc, only the F I lines were found, cf. spectra from 170 µs to about 100 µs. The line intensities rapidly faded out due to the low decreasing energy input by the arc. Then a "dark" phase followed without detectable emission starting from about 100 µs before and lasting until about 100 µs after current zero. During this dark phase, other spectral techniques would be necessary for investigation of the residual plasma, e.g., intensified cameras for emission spectroscopy or absorption measurements. The first spectra with sufficient intensity after CZ shows a different behavior: Additional lines could be observed, namely the oxygen triplet O I at 777 nm and three N I nitrogen lines around 744 nm. They are marked by red to differ them from the fluorine lines and to prove that they were not detectable before CZ. This demonstrates the back-flow of ambient air containing oxygen and nitrogen into the nozzle after CZ. The temporal evolution of three exemplary lines is shown in the left lower part of Figure 6. It should be mentioned that for this plot first the line integral of each of the three lines was calculated, second the intensity at line edges (lower/higher wavelength) was taken as "line background signal" and subtracted, and finally the intensities of these line integrals were normalized for a better comparison. The two fluorine lines show a very similar behavior, decreasing before and increasing intensity after CZ with comparable falling/rise rates. In contrast, the oxygen lines (triplet 777 nm) have an intensity of background level during all times before and until 100 µs after CZ. Then they rise up very fast, i.e., from zero to a maximum level within about 100 µs. A slower decrease follows however, they can be detected until 500–600 µs. Thus, for both the ignition phase as well as the current zero crossing, it can be stated for the high current and setup (a) that the arc plasma needs about 0.5–1.0 ms to be completely dominated by the ablated wall material and the influence of the surrounding gas can be neglected—at least under ambient conditions of one bar air.

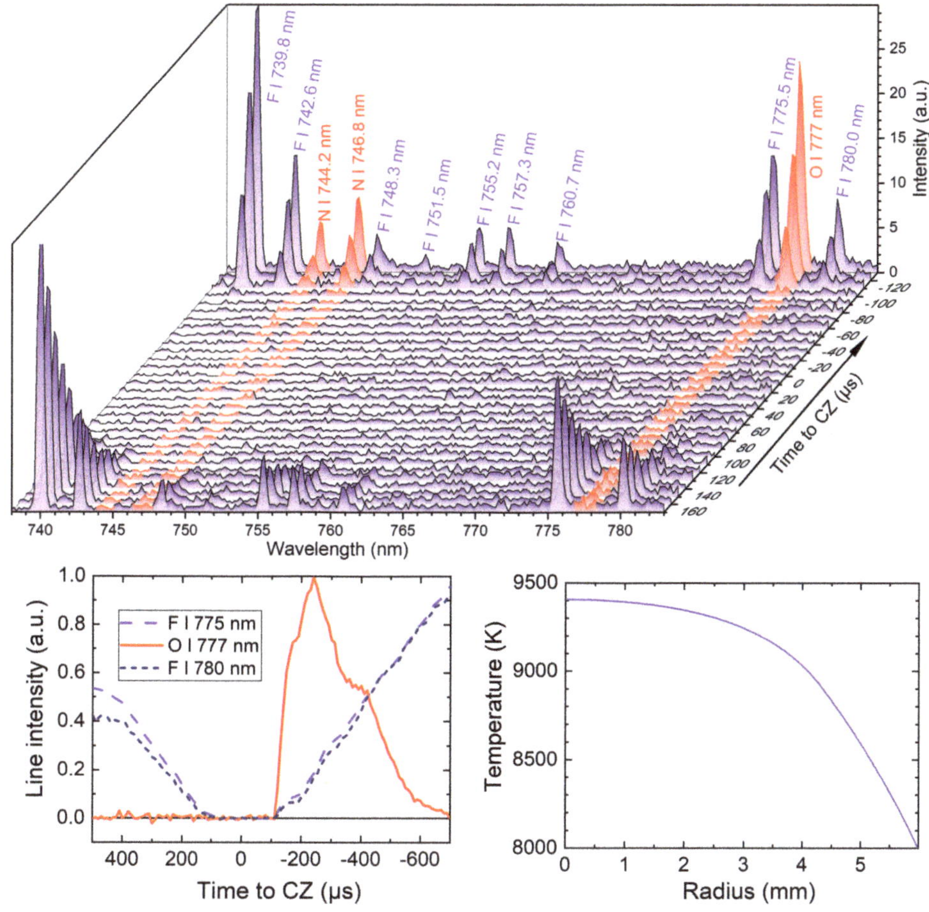

Figure 6. Top: Temporal evolution of spectra around current zero in the arc center for setup (b). Note the direction of the timescale from bottom to top. **Bottom left**: Temporal evolution of three selected line intensities. **Bottom right**: Radial temperature profile acquired 400 µs before CZ.

The next step is the determination of radial temperature profiles as close as possible to current zero with HSC. Hence, sufficient spatial resolution is required. Therefore, the number of lines used for the two-dimensional spectra was increased to 600. As a consequence, the repetition rate and thus, the temporal resolution had to be reduced to 133 µs and 7500 fps, respectively. Hence, the dark phase is reduced to one spectrum only. Furthermore, there are two spectra directly before CZ with an emission sufficient for F I line detection but not for plasma temperature determination due to restrictions in the signal-to-noise ratio. The third spectrum, i.e., 400 µs before CZ, could be applied. An absolute intensity calibration of the 2D spectrum was done by means of a tungsten ribbon lamp. Then Abel inversion was carried out to obtain radially resolved emission coefficients under the assumptions of an optically thin plasma and rotational symmetry of the arc. Finally, radial temperature profiles were determined assuming a plasma composition of 100% C_2F_4 at atmospheric pressure (no contribution from electrodes or air) and local thermodynamic equilibrium. Several fluorine lines were applied for comparison, yielding similar results. The temperature profile shown in the right part of Figure 6 has a maximum of 9400 K in the arc center and is rather broad: Even at a radial position of about 4 mm, i.e., 2 mm away from the nozzle wall, the temperature is still around 9000 K. In close vicinity to the wall, it decreases rapidly to values around 8000 K.

3.3.2. OES with Intensified Camera

The case of a current zero phase shall be investigated under the more realistic setup (b) in the model circuit chamber filled with 1 bar CO_2 (5 kA at 50 Hz). The high–speed camera is replaced by an ICCD camera (Additionally, also a video file was added as supplementary material to this article (wavelength range 774.5–781.5 nm). It shows the change from oxygen to fluorine line emission and back, but lacks of temporal resolution and intensity close to CZ). Thus, an improvement in sensitivity can be demonstrated as a side effect. An example of a two-dimensional spectrum is shown in the upper part of Figure 7. It was acquired 300 µs before CZ with an exposure time of 200 ns. The vertical axis comprises the full nozzle cross section with a diameter of 12 mm.

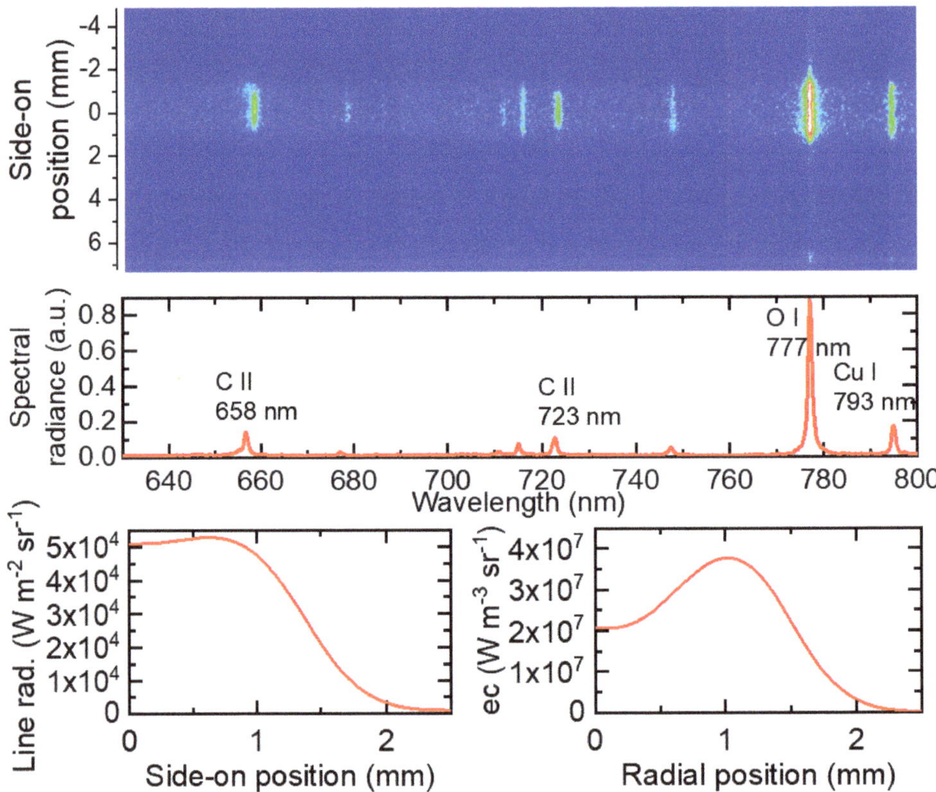

Figure 7. Two-dimensional spectrum measured 300 µs before CZ by an intensified CCD (ICCD) camera to gain higher sensitivity in comparison to HSC (**top**). Vertical axis comprises of a nozzle diameter of 12 mm. After shifting the axis to the central position of the arc and symmetrization, side-on radiance was obtained by integration over the O I 777 nm triplet (**bottom left**). The deduced emission coefficient (**bottom right**) shows a distinct maximum at 1 mm caused by the higher ionization degree in the arc center.

Obviously, the arc was contracted to a few mm due to the low current (<500 A) and was slightly shifted upwards, i.e., not in the center of the nozzle. For further analysis, the symmetry axis had to be shifted by about 1.5 mm from the nozzle center to the central position of the arc. Thus, symmetrization of the two sides ("upper" and "lower") can be carried out, providing a side-on spectrum with spatial dependence from central to outer side-on positions. The spectral radiance for the position of the arc center is given as a 1D-spectrum in the middle of Figure 7. Due to the CO_2 gas filling, atomic oxygen (O I 715 nm, 777 nm), and ionic carbon (C II 658 nm, 723 nm), lines are found but none from fluorine, e.g., F I at 775.5 nm and 780.0 nm. In general, F I lines could definitely be detected at 2 ms but never at 1 ms or less before CZ under these conditions. This will be illustrated later, cf. the collection of spectra at different times before CZ in Figure 9. This is a consequence of the different setup and lower arc current compared with the case described above (lower peak current of 5 kA instead of 11 kA, and longer duration of the half-wave of 10.7 ms instead of 5 ms), yielding less vapor from PTFE ablation. Additionally, atomic copper line emission Cu I 793.3 nm was observed.

Both the ionic carbon and atomic oxygen lines were used for the analysis of the radial temperature distribution. In the following, the procedure will be described for the oxygen triplet. At the first step, an absolute intensity calibration was carried out. Secondly, the total radiation flux of the O I 777 nm triplet was obtained by spectral integration over the line (~775–780 nm with correction using left and right background) for each side-on position and the resulting line radiance in units of $W\,m^{-2}\,sr^{-1}$ is plotted in the left bottom of Figure 7). This side-on profile was found to be rather flat in the first millimeter beginning from the central position and to decrease practically to zero within another millimeter. Although the arc center was not in the nozzle center, it could be assumed that the main path of current flow has sufficient rotational symmetry for the application of an inverse Abel transformation. Hence, in the third step the radial-dependent emission coefficient *ec* of the O I triplet could be determined as shown in the right bottom of Figure 7. A distinct maximum of *ec* was found at a radial position of $R_{max} \approx 1$ mm, towards the arc center *ec* decreased again though higher temperatures are to be expected there. Such a behavior gives hint that the "normal maximum" of the emission coefficient for a line transition is reached. This maximum results from ionization, which decreases the available number of atoms to be in an excited level with increasing temperature. For a better illustration, the temperature dependence of the emission coefficient is plotted in the left part of Figure 8. It was calculated on the base of NIST data for different pressures of pure CO_2 assuming optically thin plasma. The normal maximum was found to be around 16,000 K with only weak dependence on the pressure, see the right part of Figure 8. It should be noted that the maximum emission coefficient has nearly a linear dependence on the CO_2 pressure since it principally reflects the density of the radiating species.

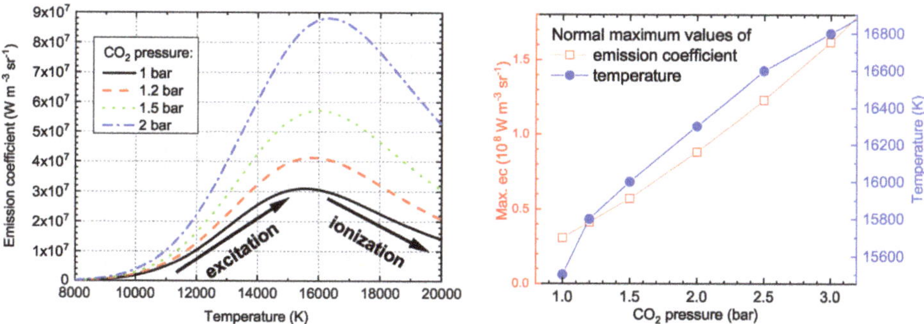

Figure 8. Left: Calculated emission coefficient of the O I 777 nm triplet at different partial pressures of CO_2 depending on temperature. Note that the normal maximum is strongly dependent on the pressure but its position is rather stable around 16,000 K. **Right**: Pressure dependence of emission coefficient and temperature at normal maximum.

As a fourth step to temperature profiles, plasma temperatures can be obtained from the experimentally determined emission coefficients by comparison, with the calculated values in Figure 8. Therefore, in the case of experimental *ec* from the "outer" part of the arc, i.e., values for radial position $r > R_{max}$, the left wing of the curve ("rising" = temperatures below normal maximum) has to be applied, while for the arc center $r < R_{max}$ the right wing is valid ("falling" = temperature above normal maximum). Knowing the pressure and plasma composition, an absolute value of the emission coefficient would not be mandatory. Hence, the absolute intensity calibration gives additional information and allows one to validate the experimental methods and assumptions. In general, the highest experimentally determined emission coefficients fitted best to the calculated normal maxima for a plasma composition of pure CO_2 with 1.2 bar total pressure. The ionic carbon line C II at 658 nm shows the normal maximum at around 22,000 K. The experimentally obtained *ec* did show no indication for temperature above the normal maximum, which proves that double ionization was rather improbable in accordance with earlier studies. Hence, for C II only the left wing with "rising" *ec* were applied. However, the experimentally obtained emission coefficients are higher than the normal maximum calculated for a pressure of 1.2 bar in pure CO_2 (the assumption used for the evaluation of the O I triplet). This is an indication that an additional amount of carbon is in the gas mixture, eventually originating from the delayed evaporation of carbon soot particles produced during nozzle ablation. Because the gas composition and pressure could not be determined accurately, pure CO_2 but with an enhanced pressure of 2 bar (close to the results of CFD simulations) was assumed for the evaluation of the *ec* of the C II line.

Spectral radiance for different instants of time before current zero are shown in Figure 9. For a better visibility, only parts of the spectral range, comprising the most relevant lines and six selected spectra are plotted. A logarithmic intensity scale is used to show the enormous decrease of line intensities, e.g., by about 2 orders of magnitude for O I and Cu I. Note that the exposure time had to be increased from 0.2–0.5 μs to 2.0 μs for all spectra closer to CZ than 150 μs due to the decrease of intensity. Fluorine atomic lines were found only in the gray spectrum acquired 2 ms before CZ, proving that under the conditions of this setup and current waveform the influence of ablated wall material was negligible ≤ 1 ms before CZ. Carbon ionic lines could be detected until 80 μs before CZ whereas atomic oxygen and copper lines were visible in all spectra.

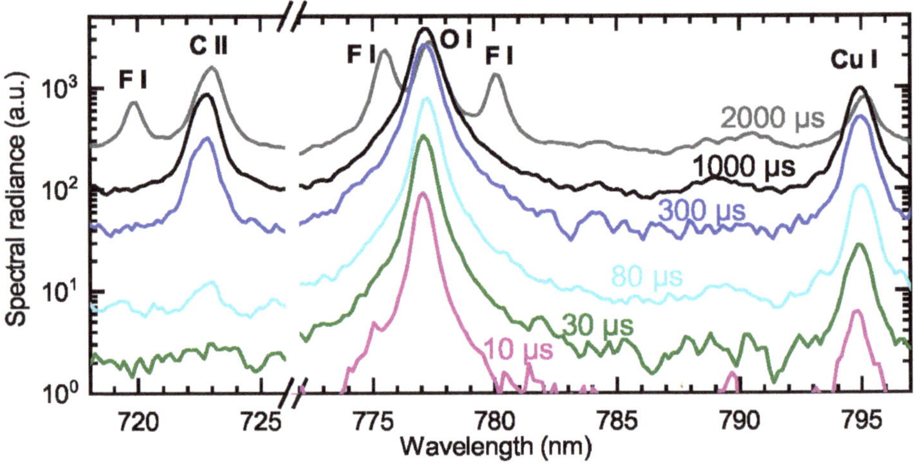

Figure 9. Selection of spectra acquired at different instants of time before the current zero.

The radial temperature profiles for five shots obtained from the emission coefficient of the C II at 658 nm considering 100% CO_2 gas at 2 bar are plotted in the upper part of Figure 10. Only times of 1 ms or less before CZ were taken into the analysis, i.e., the spectrum at 2 ms was excluded due to the F I lines indicating still PTFE admixtures. The spectrum acquired 1 ms before current zero resulted in a broad and flat temperature profile. The temperature between the arc center and 1.5 mm was more or less constant around 18,000 K. Then it was continued with a slow decrease to about 12,000 K at a radial position of nearly 4 mm, e.g., 2 mm away from the nozzle wall. This is typical behavior of a wall-stabilized arc at moderate current. For the following spectra, i.e., taken after 400 and 300 µs before CZ, two effects could be observed: (i) A considerable decrease of the arc diameter by nearly a factor of two that was accompanied by (ii) an increase of the core temperature to values above 20,000 K. Hence, a transition from the broad profile due to wall-stabilization to smaller, constricted profiles occurred. Since the arc current did not decrease as quickly as the square of the arc diameter, the core temperature had to be increased to provide a sufficient current density. In the following period, i.e., for times 130–150 µs before CZ, a further decrease of the arc diameter was found however, the maximum plasma temperature in the arc core could also be decreased due to further current decrease. Spectra acquired at less than 100 µs before CZ had too low intensities of the C II ionic line to allow for a determination of reliable temperature profiles.

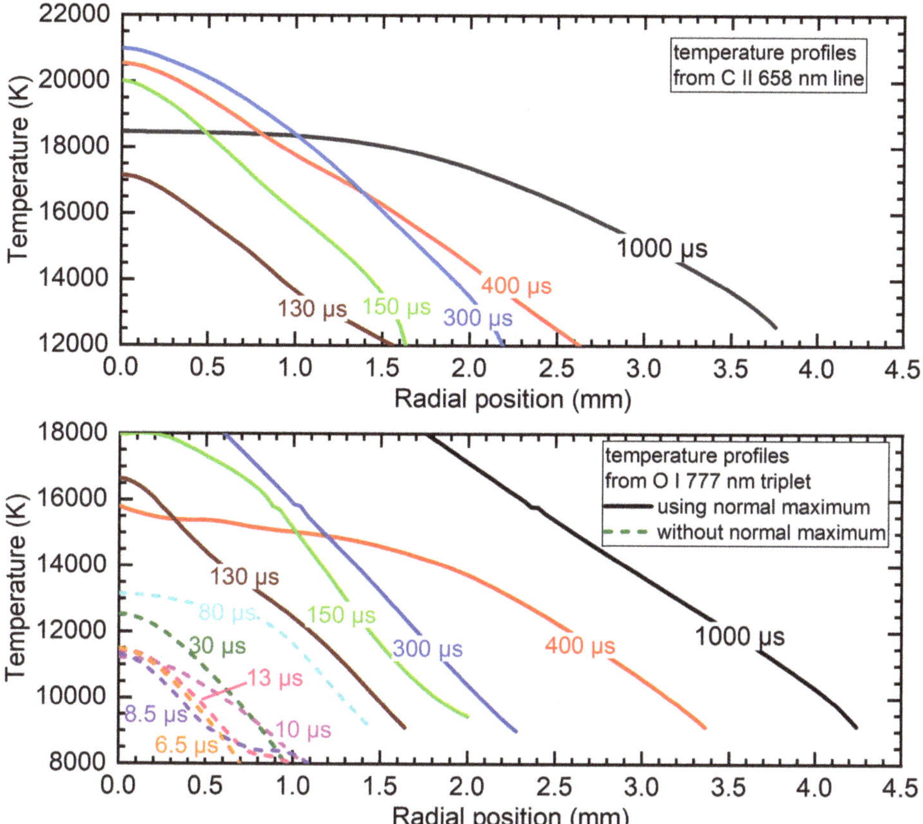

Figure 10. Radial profiles of plasma temperature obtained from ionic carbon line C II at 658 nm (**top**) and from atomic oxygen triplet at 777 nm (**bottom**) with (130–1000 µs) and without (<100 µs before CZ, dashed curves) application of normal maximum.

In the lower part of Figure 10, radial temperature profiles for numerous shots obtained from the O I triplet at 777 nm are plotted . Here, an emission coefficient for 100% CO_2 with a 1.2 bar total pressure has been considered for the temperature determination. In agreement with the results for the C II line, the broadest profile with the highest temperatures was obtained for 1 ms. Here, the position of the normal maximum connected with a fairly established temperature of 16,000 K was obtained at about 2.5 mm. Although the maximum temperature exceeded 20,000 K only values till 18,000 K were plotted due to higher uncertainties for the central position. However, the plasma temperature was still about 10,000 K even at a radial position of 4 mm, i.e., the arc was spread over most of the nozzle area (12 mm diameter). For the following profiles, the effect of temperature increase in the core was not as clear as above for C II. Though some profiles do not perfectly fit into the generally smooth decreasing behavior, the general tendency was that both arc diameter and maximum temperature in the center decreased continuously with the decreasing current. The above described method based on the normal maximum of the oxygen triplet could be applied to all spectra taken ∼100–1000 µs before CZ. For the remaining spectra (<100 µs) the experimental emission coefficients were below the normal maximum. Hence, the common single-line method was applied using only the left ("rising") wing of the curve from Figure 8. These temperature profiles are plotted as dashed curves in Figure 10. The tendency to smaller profiles with lower peak temperatures can be followed until about 10 µs before current zero, i.e., at very low currents of few tens of amperes. Although spectra (of 2 µs exposure time) were taken at varying times of 13, 10, 8.5, and 6.5 µs before CZ and the exact acquisition times were measured by a comparison of current waveform and a monitor signal from the ICCD camera, nearly the same results were obtained for all four shots: A temperature maximum slightly above 11,000 K and a decrease to 8000 K within ≤1.0 mm. Here, the experimental limit is reached with possible changes in the temperature profile superimposed by shot-to-shot variation of the discharge itself, window transmission, and uncertainties in determination of the emission coefficient. As a control, another spectrum was acquired at current zero (from 1.0 µs before CZ to 1.0 µs after CZ) however, only the noise was recorded.

4. Discussion

4.1. General

An ablation-dominated arc of 5 kA peak current was operated in a model circuit breaker with a CO_2 atmosphere. The application of a slit over the full radius of the PTFE nozzle enabled a direct investigation inside the nozzle. This was substantial progress in comparison with previous measurements that potentially suffered from an influence of turbulent gas flow in the heating channel [17]. Sealing by thin quartz plates proved to be a useful method to obtain reproducible conditions of discharges without significant changes of material flow or plasma conditions. Consequent exchange of the plates after each shot yielded high window transmission with moderate blackening. An averaged overall transmission of about 50% was estimated along the observation axis made of the quartz plate in the nozzle and windows of the MCB chamber as well as the high-pressure vessel. The shot-to-shot variation was usually about 10 %. Nozzle ablation caused a widening of the nozzle diameter and thus a slow reduction of maximum pressures over many shots. Therefore, the PTFE nozzles were exchanged regularly. Only very few cases of reduced sealing quality occurred, easily noticeable after discharges due to increased blackening at the plate's corners. These shots had be repeated with renewed sealing.

The assets of different methods for optical investigation were demonstrated. Using high-speed cameras the general arc behavior was investigated, e.g., revealing rotational symmetry over the whole nozzle diameter in the high-current phase but not close to current zero when the stabilization by ablation of nozzle wall was lost and the arc was constricted and out of the nozzle center. Besides, the HSC allowed an investigation of many successive points of time within the same shot. Thus, reproducibility was not demanded and dynamic changes, their rising and falling times, and shot-to-shot variation could be easily investigated. Although parameter variation was not the focus of this work, such investigations are rather comfortable using this technique. Furthermore, a combination of HSC with double frame optics was introduced, filtering one channel only for radiation from CO_2 (O I at 777 nm) and the other channel only for emission from PTFE (F I at 675 nm), which allowed us to gain knowledge about the temporal evolution of the plasma composition. In combination with according CFD simulation, gas flow behavior could be analyzed, including the exact determination of the point in time when flow reversal occurred before current zero. In the experiments described here, only qualitative analysis could be carried out. For a quantitative description, more knowledge about plasma composition and absolute intensity calibration would be mandatory [23].

4.2. OES Using HSC

Deeper information was obtained from spatially- and temporally-resolved video spectroscopy using HSC. That comprised of the different phases of discharge and the occurrence of Swan band emission from C_2 molecules that are treated in an accompanying publication [20]. Radial temperature profiles have been determined until 400 µs before current zero. Assuming a plasma composition of 100% C_2F_4 at atmospheric pressure, a broad temperature profile has been obtained with a maximum of 9400 K in the arc center and about 9000 K at a radial position of about 4 mm, i.e., 2 mm away from the nozzle wall. Several fluorine lines were applied for comparison, yielding similar results. A "dark window" without detectable emission was observed due to low intensities caused by cold gas flow and low current on the one hand and limitations in the sensitivity of high-speed cameras on the other hand, starting in a best case about 100 µs before CZ. Furthermore, it could be stated for the high current and setup (a) that the arc plasma needs about 0.5–1.0 ms for both the ignition phase as well as the current zero crossing to be completely dominated by the ablated wall material. The influence of the surrounding gas can be neglected, at least under ambient conditions of one bar air.

4.3. OES with ICCD

The sensitivity of OES was increased by application of OES with an intensified CCD camera, allowing single-shot measurements until a few µs before current zero. Two lines were used for the determination of temperature profiles of the arc plasma, whereas the ionic carbon line C II at 658 nm has a normal maximum around 22,000 K and therefore better sensitivity concerning temperatures around 18,000–20,000 K, the atomic oxygen triplet O I at 777 nm has its normal maximum around 16,000 K and higher sensitivity at lower temperatures. Off-axis maxima of the radial emission coefficient of the O I triplet were found, indicating temperatures in the arc center above and in the arc fringes below the normal maximum. Hence, the normal maximum can be used for the calibration of the emission coefficient according to the Fowler–Milne method. In addition, the absolute intensity calibration by a radiation normal has been used for verification. The emission coefficient of the C II line was evaluated with absolute intensity calibration only.

As experimental uncertainties of the determined emission coefficients, in particular the window transmission (estimated to 50% ± 10%), adjustment of slit width (50 ± 5 µm), performing absolute intensity calibration (uncertainty up to 20%), and pressure measurement (uncertainty up to 10%) have to be considered. It is an advantage of the applied method for the O I triplet that the normal maximum is independent from influences by transmission, absorption, and absolute intensity calibration. Therefore, the uncertainty of the temperature determined from O I is low around the normal maximum of 16,000 K as well as in the range of 10,000–14,000 K (up to 10%) due to the exponential intensity rise with temperature. However, reliable temperatures above 20,000 K cannot be determined from the O I triplet. An important factor is the remaining uncertainty of the gas composition and the partial pressure of carbon and oxygen. Pressure measurements showed a shot-to-shot variation between 0.2 and 0.4 bar, and the fitting of the emission coefficients of the C II line and the O I triplet with respect to the corresponding normal maximum in pure CO_2 lead to the different pressures 1.2 and 2 bar. But considering the relatively low variation of the temperatures at the normal maxima with pressure, the uncertainty of the plasma temperatures at least around the normal maximum is below ±200 K. For times closer than 100 µs to CZ, where the emission coefficient of the O I triplet could be evaluated by absolute intensity calibration only, the uncertainty depends on the rise of line emission coefficient with temperature. For temperatures below 11,000 K, the emission coefficient is possibly underestimated by factor 2 in maximum, which causes an underestimation of temperature by 900 K.

It has been found that the arc was rather broad at 1000 μs and 400 μs before CZ in agreement with the observations of wall-stabilization. In the following, arc constriction was observed. Despite the uncertainties discussed above, it can be stated that the maximum temperature decreased from above 18,000 K at 300 μs to about 11,000 K at 10 μs before CZ.

5. Conclusions

The main challenge concerning optical investigations of ablation-dominated high-current arcs close to current zero is a low intensity of the emitted radiation. Therefore, different techniques of high-speed camera (HSC) imaging with or without filtering and optical emission spectroscopy (OES) were introduced and their assets and limitations were discussed. It was shown that some important effects can be analyzed even with rather simple nozzle experiments, applicable to other groups and setups.

Two setups were used: With the first setup, experiments with a long, tubular nozzle were applied on to study the CZ transition including new re-ignition of the arc. Using OES with HSC, a dark period of 200 μs around CZ was observed and the differences in spectra before and after CZ were discussed. Radial temperature profiles could be obtained until 400 μs before CZ. Here, a typical case of a well-established arc was found with a broad temperature profile, characteristic for an arc stabilized by ablation of the nozzle wall.

For the second setup, a model circuit chamber in CO_2 atmosphere was used to study more realistic, praxis-relevant features including flow reversal, arc behavior around CZ, and arc before vanishing. Using OES with an intensified CCD camera, higher sensitivity was realized allowing the determination of temperature profiles. Whereas ionic carbon lines were applied mainly for quantitative characterization at higher temperatures until 100 μs, atomic oxygen lines delivered quantitative profiles until a few microseconds before CZ with a higher sensitivity at lower temperatures. Generally, a transition was observed in the arc behavior. Until several hundred microseconds before CZ, the arc was wall-stabilized with a broad and rather flat temperature profile. After vanishing of wall stabilization and inflow of cold gas, a highly dynamic arc appeared that was constricted and asymmetric moving out of center. During the transition, the maximum temperatures in the core increased to yield higher current densities for the constricted arc.

In future work it will be important to combine the experimental results and modeling concerning temperature profiles, composition calculation for a determination of current density.

Supplementary Materials: The following are available online at http://www.mdpi.com/1996-1073/13/18/4714/s1, Video S1 (DFO-fig4.mp4) : High–speed imaging using double frame optics according to Figure 4. Video S2 (OES-fig5.mp4):Video S2: High–speed imaging optical emission spectroscopy according to Figure 5. Video S3 (OES-fig7.mp4):Video S3: High–speed imaging optical emission spectroscopy around the O I triplet at 777 nm with higher spectral resolution according to Figure 7.

Author Contributions: Conceptualization, R.M. and D.U.; methodology, validation, and formal analysis, R.M.; investigation with setup (a), R.M. and A.K.; investigation with setup (b), R.M., A.K., and N.G.; writing—original draft preparation, R.M. and D.U.; project administration, D.U. All authors have read and agreed to the published version of the manuscript.

Funding: This research was funded by Deutsche Forschungsgemeinschaft grant numbers UH 106/13-1 and SCHN 728/16-1.

Acknowledgments: The authors would like to thank Steffen Franke for experimental help and fruitful discussions. The calculation of plasma composition was realized by Sergey Gortschakow (all Leibniz Institute for Plasma Science and Technology).

Conflicts of Interest: The authors declare no conflict of interest. The funders had no role in the design of the study; in the collection, analyses, or interpretation of data; in the writing of the manuscript, or in the decision to publish the results.

Abbreviations

The following abbreviations are used in this manuscript:

CB	Circuit Breaker
CO_2	carbon dioxide
CFD	computational fluid dynamics
CZ	current zero
DFO	double frame optics
ICCD	intensified charge coupled device
HSC	high-speed camera
LTE	local thermal equilibrium
MoS_2	molybdenum disulfide
OAS	optical absorption spectroscopy
OES	optical emission spectroscopy
PTFE	polytetrafluoroethylene
PMMA	polymethyl methacrylate
SF_6	sulfur hexafluoride
W–Cu	tungsten–copper

References

1. Seeger, M. Future Perspectives on High Voltage Circuit Breaker Research. *Plasma Phys. Technol.* **2015**, *2*, 271–279.
2. Seeger, M. Perspectives on research on high voltage gas circuit breakers. *Plasma Chem. Plasma Process.* **2015**, *35*, 527–541. [CrossRef]
3. Glaubitz, P.; Stangherlin, S.; Biasse, J.; Meyer, F.; Dallet, M.; Pruefert, M.; Kurte, R.; Saida, T.; Uehara, K.; Prieur, P.; et al. CIGRE position paper on the application of SF_6 in transmission and distribution networks. *Electra* **2014**, *34*, 34–39.
4. Christophorou, L.; Olthoff, J.K.; Green, D.S. *Gases for Electrical Insulation and Arc Interruption: Possible Present and Future Alternatives to Pure SF_6;* NIST Technical Note 1425; Nov. 1997; pp. 1–44. Available online: https://www.nist.gov/publications/gases-electrical-insulation-and-arc-interruption-possible-present-and-future (accessed on 9 September 2020).
5. Robin-Jouan, P.; Bousoltane, K.; Kieffel, Y.; Trepanier, J.Y.; Camarero, R.; Arabi, S.; Pernaudat, G. Analysis of last development results for high voltage circuit-breakers using new g^3 gas. *Plasma Phys. Technol.* **2000**, *33*, 2583–2590. [CrossRef]
6. Godin, D.; Trepanier, J.Y.; Reggio, M.; Zhang, X.D.; Camarero, R. Modelling and simulation of nozzle ablation in high-voltage circuit-breakers. *J. Phys. D Appl. Phys.* **2000**, *33*, 2583–2590. [CrossRef]
7. Wu, Y.; Wang, C.; Sun, H.; Murphy, A.B.; Rong, M.; Yang, F.; Chen, Z.; Niu, C.; Wang, X. Properties of C_4F_7N–CO_2 thermal plasmas: Thermodynamic properties, transport coefficients and emission coefficients. *J. Phys. D Appl. Phys.* **2007**, *51*, 155206. [CrossRef]
8. Seeger, M.; Smeets, R.; Yan, J.; Ito, H.; Claessens, M.; Dullni, E.; Falkingham, L.; Franck, C.M.; Gentils, F.; Hartmann, W.; et al. Recent Trends in Development of High Voltage Circuit Breakers with SF_6 Alternative Gases. *Plasma Phys. Technol.* **2017**, *4*, 8–12. [CrossRef]
9. Stoller, P.C.; Seeger, M.; Iordanidis, A.A.; Naidis, G.V. CO_2 as an arc interruption medium in gas circuit breakers. *IEEE Trans. Plasma Sci.* **2013**, *41*, 2359–2369. [CrossRef]
10. Guo, Z.; Liu, S.; Pu, Y.; Zhang, B.; Li, X.; Tang, F.; Lv, Q.; Jia, S. Study of the arc interruption performance of CO_2 gas in high-voltage circuit breaker. *IEEE Trans. Plasma Sci.* **2019**, *47*, 2742–2751. [CrossRef]
11. Kozakov, R.; Kettlitz, M.; Weltmann, K.D.; Steffens, A.; Franck, C.M. Temperature profiles of an ablation controlled arc in PTFE: I. Spectroscopic measurements. *J. Phys. D Appl. Phys.* **2007**, *40*, 2499–2506. [CrossRef]
12. Methling, R.; Franke, S.; Uhrlandt, D.; Gorchakov, S.; Reichert, F.; Petchanka, A. Spectroscopic Study of Arc Temperature Profiles of a Switching-off Process in a Model Chamber. *Plasma Phys. Technol.* **2015**, *2*, 163–166.
13. Ruchti, C.B.; Niemeyer, L. Ablation controlled arcs. *IEEE Trans. Plasma Sci.* **1986**, *14*, 423–434. [CrossRef]
14. Bort, L.; Schultz, T.; Franck, C.F. Determining the time constant of arcs at arbitrary current levels. *Plasma Phys. Technol.* **2019**, *5*, 175–179. [CrossRef]

15. Schultz, T.; Hammerich, B.; Bort, L.; Franck, C.F. Improving interruption performance of mechanical circuit breakers by controlling pre-current-zero wave shape. *High Volt.* **2019**, *4*, 122–129. [CrossRef]
16. Methling, R.; Khakpour, A.; Wetzeler, S.; Uhrlandt, D. Investigation of an ablation–dominated arc in a model chamber by optical emission spectroscopy. *Plasma Phys. Technol.* **2017**, *4*, 153–156. [CrossRef]
17. Eichhoff, D.; Kurz, A.; Kozakov, R.; Gött, G.; Uhrlandt, D.; Schnettler, A. Study of an ablation–dominated arc in a model circuit chamber. *J. Phys. D Appl. Phys.* **2012**, *45*, 305204. [CrossRef]
18. Seeger, M.; Tepper, J.; Christen, T.; Abrahamson, J. Experimental study on PTFE ablation in high voltage circuit-breakers. *J. Phys. D Appl. Phys.* **2006**, *39*, 5016–5024. [CrossRef]
19. Seeger, M.; Niemeyer, L.; Christen, T.; Schwinne, M.; Dommerque, R. An integral arc model for ablation controlled arcs based on CFD simulations. *J. Phys. D Appl. Phys.* **2006**, *39*, 2180–2191. [CrossRef]
20. Methling, R.; Götte, N.; Uhrlandt, D. Ablation-Dominated Arcs in CO_2 atmosphere—Part II: Molecule emission and absorption. *Preprints* **2020**. [CrossRef]
21. Kramida, A.; Ralchenko, Y.; Reader, J.; Team, N.A. *NIST Atomic Spectra Database (Version 5.7.1)*; National Institute of Standards and Technology: Gaithersburg, MD, USA, 2019. Available online: http://physics.nist.gov/asd (accessed on 9 September 2020).
22. Kurucz, R.L.; Bell, B. Atomic Line Data. Kurucz CD-ROM No. 23. Smithsonian Astrophyical Observatory. 1995. Available online: http://www.cfa.harvard.edu/amp/ampdata/kurucz23/sekur.html (accessed on 9 September 2020).
23. Methling, R.; Franke, S.; Uhrlandt, D.; Gortschakow, S.; Panousis, E. Spectroscopic measurements of arc temperatures in a model HV circuit breaker. *Plasma Phys. Technol.* **2015**, *2*, 167–170.

Article

Ablation-Dominated Arcs in CO_2 Atmosphere—Part II: Molecule Emission and Absorption

Ralf Methling [1,*,†], Nicolas Götte [2,†] and Dirk Uhrlandt [1,†]

1 Leibniz Institute for Plasma Science and Technology (INP), Felix-Hausdorff-Str. 2, 17489 Greifswald, Germany; uhrlandt@inp-greifswald.de

2 Institute for High Voltage Technology, RWTH Aachen University, Schinkelstrasse 2, 52056 Aachen, Germany; goette@ifht.rwth-aachen.de

* Correspondence: methling@inp-greifswald.de; Tel.: +49-3834-554-3840

† These authors contributed equally to this work.

Received: 11 August 2020; Accepted: 9 September 2020; Published: 10 September 2020

Abstract: Molecule radiation can be used as a tool to study colder regions in switching arc plasmas like arc fringes in contact to walls and ranges around current zero (CZ). This is demonstrated in the present study for the first time for the case of ablation-dominated high-current arcs as key elements of self-blast circuit breakers. The arc in a model circuit breaker (MCB) in CO_2 with and an arc in a long nozzle under ambient conditions with peak currents between 5 and 10 kA were studied by emission and absorption spectroscopy in the visible spectral range. The nozzle material was polytetrafluoroethylene (PTFE) in both cases. Imaging spectroscopy was carried out either with high-speed cameras or with intensified CCD cameras. A pulsed high-intensity Xe lamp was applied as a background radiator for the broad-band absorption spectroscopy. Emission of Swan bands from carbon dimers was observed at the edge of nozzles only or across the whole nozzle radius with highest intensity in the arc center, depending on current and nozzle geometry. Furthermore, absorption of C_2 Swan bands and CuF bands were found with the arc plasma serving as background radiator. After CZ, only CuF was detected in absorption experiments.

Keywords: circuit breaker; switching arc; optical emission spectroscopy; optical absorption spectroscopy; current zero; SF6 alternative gases; CO_2; PTFE; Swan bands; CuF

1. Introduction

Self-blast circuit breakers represent one of the main technologies for high-current interruption at high voltage. After contact separation, intense radiation emitted from the high-current arc leads to a considerable photo-ablation of the surrounding nozzle which causes a pressure buildup and finally a strong gas flow necessary for arc quenching around current zero (CZ) [1,2]. Polytetrafluoroethylene (PTFE) is typically used as the nozzle material and SF_6 as the filling gas. However, the substitution of the greenhouse gas SF_6 by more environmentally-friendly gases like CO_2 is an actual trend. The pressure buildup due to strong arc radiation and nozzle ablation, as well as the arc quenching processes, are key issues of the successful current breaking and have been subject to a large number of scientific studies. The main questions concern the properties of the arc and the hot gas regions like temperatures and species densities which are required for a sufficient understanding of the processes. Optical methods, like emission and absorption spectroscopy, can provide such quantities under the demand that arc and hot gas regions are optically accessible. However, an optical access can only be realized by adapted construction of specific model circuit breakers (MCB) or by appropriate model experiments [3–5].

Meanwhile, a sufficiently good knowledge of the arc properties during the high-current phase and in the high-temperature regions (above 6000 K) of the arc has been developed from spectroscopic studies of switching arc experiments and MCBs (see e.g., [3]). This is because atomic and ionic species dominate in the high-temperature regions and generate an intense spectral line radiation which can be well used for the determination of temperature and species densities [6–8]. However, the analysis of low-temperature regions of the arc fringes, of the regions near nozzle walls and of the temporal phase of arc quenching is much more challenging due to low line radiation intensities.

The investigation of the phase around current zero by optical emission spectroscopy (OES) and the determination of arc temperatures during the arc quenching as close as possible to CZ was a topic of our accompanying paper [9]. An MCB using CO_2 as a filling gas and a PTFE-nozzle experiment under ambient air were used for the analysis of line radiation of oxygen and fluorine atoms as well as of carbon ions. Both setups will also be used in this study and explained shortly in Section 2.

It is well-known from composition calculations of thermal plasmas that the dissociation of filling gases like SF_6 and CO_2, reactions with the ablation product C_2F_4 and metal vapor from electrode erosion can produce a number of molecular species in an intermediate temperature range before an almost complete dissociation of atoms occurs at higher temperatures (see e.g., [10]). Mixtures of CO_2 with higher amount of C_2F_4 are expected to contain considerable amounts of molecules at temperatures above 3000 K, namely CF_4, CF_3, CF_2, C_2F, C_3, C_2, CF, and CO (in order of dissociation with increasing temperatures) [11]. Hence, the study of molecule radiation can help to analyze the interesting ranges of lower temperatures near the nozzle boundaries and in the arc quenching phases. Unfortunately, there is a very low number of such studies for arcs in corresponding gas mixtures and particularly for switching arcs.

Interesting candidates for the study of molecule radiation are the Swan bands of the C_2 molecule (around 500 nm) or the violet band of CN (around 385 nm) because of the relatively intense radiation in the optical range. Emission and absorption spectroscopy of the C_2 radiation have been used for example to study the structure of carbon arcs for nanoparticle synthesis [12–14]. The radiation of CN was analyzed in a study of the arc ablation of organic materials in ambient air with close relation to low-voltage switching [15]. Furthermore, both molecules have been more intensely studied in plasmas produced by laser ablation or in the laser-induced breakdown [16–20].

The occurrence of C_2 molecules is expected in switching arcs in CO_2 atmosphere or in the case of ablation of PTFE or organic wall materials. However, most of the recent research on Swan bands C_2 was carried out by laser-induced breakdown spectroscopy. In case of lower laser irradiance, the production of C_2 molecules is dominated by excitation of larger molecules like C_3, C_4 with electrons followed by photo-defragmentation, delivering exited C_2 molecules. In case of higher power, excitation resulting from electron–ion and ion–ion recombination dominates [16]. The intensity distribution of the emission pattern varies depending on pressure and temperature. Thus, an estimation of the vibrational temperature can be realized by comparison of measured and simulated spectra [17,19]. Temperatures in a thermal argon plasma interacting with various insulating plastic materials at magnetically-forced arc movement [21] and temperature decay of thermal plasmas caused by polymer ablation using inductively coupled plasma irradiation [22] were investigated experimentally and numerically. As an example of a switching arc study, the absorption spectrum of the C_2 Swan bands was analyzed in a low-voltage circuit breaker model [23]. An arc moving between polyethylene walls was considered, and the density and the rotational temperature of the C_2 molecules were determined from the absorption spectrum, which indicates the ablation of the plastic walls. Reports on the analysis of molecule radiation, the C_2 Swan bands in particular, in high-voltage switching experiments as representative for high-voltage circuit breakers are missing so far.

During the OES study of an MCB and a nozzle experiment described in our first paper [9], molecule radiation of C_2 and CuF was recorded under different conditions and to some extent in unexpected ranges of the arc. The occurrence of strong temperature gradients in the arc are already known as well as a number of molecules that might be expected to appear favorably at lower temperatures, i.e., either in the vicinity to the nozzle walls or at low currents. However, the occurrence of molecular species has been described by theoretical models (see, e.g., [11]), with a lack of experimental confirmation in many cases. In the present paper, it should be shown that some molecules are detectable under strongly varying conditions. The molecule CuF is expected when copper vapor from the electrode erosion is mixed with the dissociated PTFE vapor from the nozzle ablation [10]. The results for molecule emission and absorption should be given in this second paper in detail. The aim is to demonstrate the occurrence of molecule radiation as a possible candidate to characterize low-temperature regions in self-blast circuit breakers as well as ablation processes. However, the determination of quantities like rotational temperatures and densities is out of the scope of the present paper. The MCB and nozzle experiment setups will be presented shortly in Section 2 together with the setup for spectroscopic measurements because details can be found in [9]. Results are given in Section 3 followed by a discussion in Section 4.

2. Materials and Methods

Two setups of electrodes and nozzles were used. They are described in detail in an accompanying paper [9]; basic features are sketched in Figure 1. Actually, the majority of experiments described in this paper were carried out with setup (b) and only a few with setup (a). The electrodes were made of W–Cu with a 10 mm diameter and had a fixed distance of 40 mm. Nozzles made of PTFE doped with <0.5 wt% molybdenum disulfide (MoS_2) with an inner diameter of 12 mm were placed around the electrodes: Either setup (a) was applied with one 126 mm long, tubular-shaped nozzle of 50 mm outer diameter for strong ablation and high pressure built-up or setup (b) was used with two nozzles of about 50 mm length and 104 mm outer diameter separated by 4 mm distance to form a heating channel. At the electrode positions, the nozzle diameter was increased to about 16 mm for an exhaust gas flow.

The arcs were operated either under ambient conditions (setup (a)) or in a vessel filled with 1 bar CO_2 (setup (b)) as part of a model circuit breaker similar to [3]. Windows in both the model chamber and the vessel allowed a free view through the nozzle and hence absorption experiments.

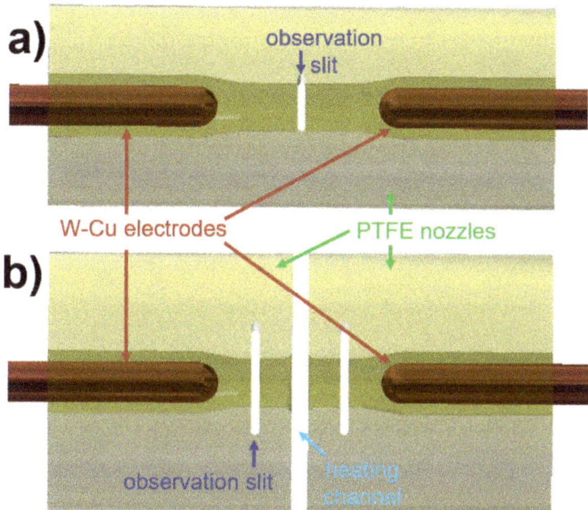

Figure 1. Setups (**a**) with a closed, long polytetrafluoroethylene (PTFE) nozzle for experiments with strong ablation and high pressure built-up and vertical observation slits in the middle and (**b**) with two separated PTFE nozzles forming a heating channel for plasma flow into a heating volume as used for the model circuit breaker.

Sine-like currents were applied for setup (a) with about 100 Hz frequency and 11 kA peak current. For setup (b) with 50 Hz and 5.3 kA. Thin Cu wires were used to initiate the arc discharges.

Currents were measured using Rogowski coils. In case of setup (b), a pressure sensor (603 A from Kistler) was placed in the heating volume of the model circuit breaker.

Optical access was realized by vertical slits of 2 mm width that were sealed by 2 mm-thick quartz plates, ranging over the complete nozzle diameter. After each shot, the sealing plates were checked visually and exchanged; the transmission was measured regularly. Pairwise placement at opposite positions enabled background illumination and absorption measurements. For setup (a) the observation slits were placed in the middle between both electrodes. In setup (b), the observation point was positioned in one of the nozzles at half distance between electrode tip and nozzle exhaust, i.e., ~9 mm away from both.

Different methods were applied for the optical analysis. Firstly, high-speed cameras (HSC) from Integrated Design Tools (IDT) were used to observe the general discharge behavior: Y6 with 24-bits color or Y4 with 10-bits monochrome. Secondly, optical emission spectroscopy was carried out by means of an imaging spectrograph with 0.5 m focal length (Roper Acton SpectraPro SP2500i). The nozzle slit was imaged on the entrance slit of the spectrograph to spectrally investigate arc cross sections, i.e., perpendicular to the arc axis. Using the spectrograph with Y4 HSC enabled to record series of 2D-spectra with typical repetition rates of 100 μs (frame rate 10 kfps), allowing rather long exposure times up to 98 μs that were necessary due to limited camera sensitivity. Alternatively, the HSC could be replaced by an intensified CCD camera (PI-MAX4 from Princeton Instruments) with higher sensitivity, allowing single frame acquisition of shorter exposure times even at lower intensities, e.g., around current zero. In a compromise between light intensity, spectral resolution, and exposure time, the entrance slit of the spectrograph was set to 50 μm. With gratings of 150 lines per mm for overview and 1800 L/mm for detailed spectra, the spectral range was 150 nm and 10 nm and the spectral resolution 0.3 nm and <0.1 nm, respectively. The intensity of side-on spectra was calibrated in units of spectral radiance by means of a tungsten strip lamp (OSRAM Wi 17/G) at the arc position. The window transmission of 50–70% was taken into account, mainly resulting from the coating of the quartz plates at the nozzles.

Thirdly, broadband absorption spectroscopy was carried out around CZ. Therefore, a background illumination was required with radiances higher or comparable to the emission of the arc. It was supplied by a pulsed high-intensity xenon lamp with a radiance similar to a Planckian radiator of 12,000 K [24]. The square-shaped pulse had about 1 ms-width at about 1 MW electric power, delivering a nearly constant emission intensity during the plateau phase.

Figure 2 shows exemplary current waveforms of the arc discharge around current zero (top, offset after CZ is caused by the Rogowski coil) and the quasi-rectangular pulsed current of the xenon lamp (bottom, red) as well as the spectrally integrated intensity measured by video spectroscopy (spectral range 400–800 nm). Since the electric pulse feeding the Xe-lamp was not perfectly rectangular, a heating phase of the xenon lamp could be observed. Thus, several Xe atomic lines were found in the first 100–200 μs of the 1 ms-pulse before a transition towards the 12,000 K-continuum emission. Additionally, with decreasing current also the emission intensity decreased. Hence, for the OAS analysis only the lamp's plateau phase was applied with a duration of about 700 μs. This relatively long, stable phase allowed for temporal investigation of absorption, e.g., compared to Z-pinches with some 10 μs of varying radiation intensity as used in [23].

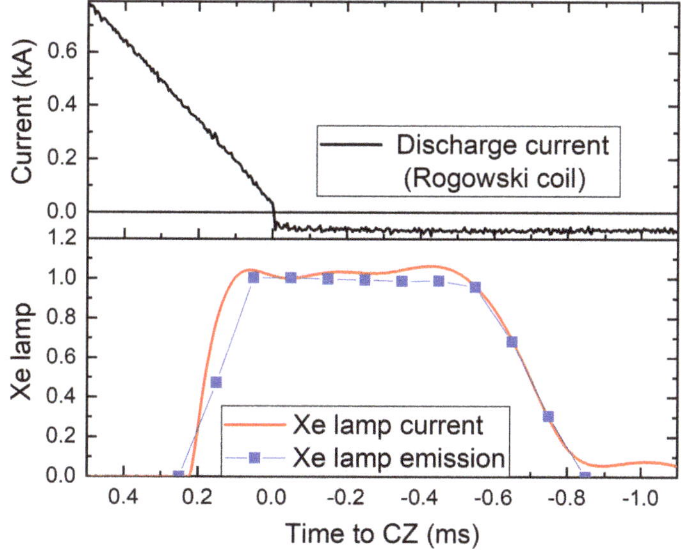

Figure 2. (**top**) Arc current at the end of discharge. (**bottom**) Xe lamp current (red) and development of its emission intensity (blue squares).

3. Results

In order to avoid doubling, some more general results that were already described and discussed in the accompanying paper will not be repeated here. That relates to electrical waveforms, temporal evolution of pressure and plasma composition in the nozzle, and video observation by HSC with and without filtering. Additionally, only selected moments from the overview video spectroscopy were shown that are mandatory for the discussion of molecule emission and absorption. It should be noted that, for easier comparison, all points in time are given with respect to the current zero crossing.

First, experiments in the MCB (setup (b)) with the sine-like current up to 5.3 kA are considered. The arc voltage was around 200 V (after peak caused by the explosion of ignition wire) until the arc extinction peak some hundred µs before CZ. The total pressure in the nozzle started from a filling pressure of 1.0 bar to a maximum of 3.5 bar close to peak current and decreased to about 2.0 bar at current zero. After ignition, an arc discharge in CO_2 atmosphere was observed, also containing copper from ignition wire and electrodes. Within the next few hundreds of microseconds, the ablation of the PTFE (C_2F_4) wall material started to dominate the discharge, blowing the CO_2 out of the nozzle. In the following, a long and stable period was observed that was dominated by ablation. Another reversal of flow was found about 2 ms before CZ: With decreasing arc current, the wall ablation and thus the pressure in the nozzle decreased to values below that in the heating volume. Hence, relatively cold gas from the heating volume with a high fraction of CO_2 flowed back into the nozzle. In the last ms, only emission from O I was observed, indicating a plasma composition completely dominated by CO_2.

3.1. *Analysis of C_2 Swan Bands*

An example of a two-dimensional spectrum is shown in Figure 3. It was acquired with setup (b) shortly before peak current (7.3 ms to CZ). On the left side, an image of the HSC observation area (grey scale image) including the OES axis (yellow dashed line) is shown. The vertical axis represents the position along the observation slit in the nozzle, cf. dashed yellow line in the HSC image on the left side; the horizontal dimension is given by the wavelength in the spectral range ~480–625 nm. The arc discharge was dominated by wall-ablation at that point in time; no emission from copper or oxygen but lines from atomic fluorine F I and atomic and ionic carbon lines C I, C II could be observed. This radiation was mainly emitted in a broad distribution over the arc cross-section with the highest intensities in central positions, as it is typical for the wall-stabilized arcs with broad and flat temperature profile [9]. However, an additional structure can be recognized with a different lateral distribution: A dense pattern of lines with increasing intensities and numbers towards higher wavelengths with abrupt breaks at positions near 516 and 564 nm, spread over the whole nozzle diameter and partly even with maxima close to the wall. This structure has been attributed to the Swan band system originating from transitions between the electronic states $d^3\,\Pi_g$ and $a^3\,\Pi_u$. Four cases of appearance of Swan bands in the discharge will be presented in the following.

Firstly, the Swan bands occurred at the outer edges of the arc preferably close to the nozzle walls as shown in Figure 3. Generally, this can be regarded as typical behavior for cases of moderate PTFE influence, i.e., when current density is not too high and the temperature close to the wall is rather low, allowing the existence of carbon dimers.

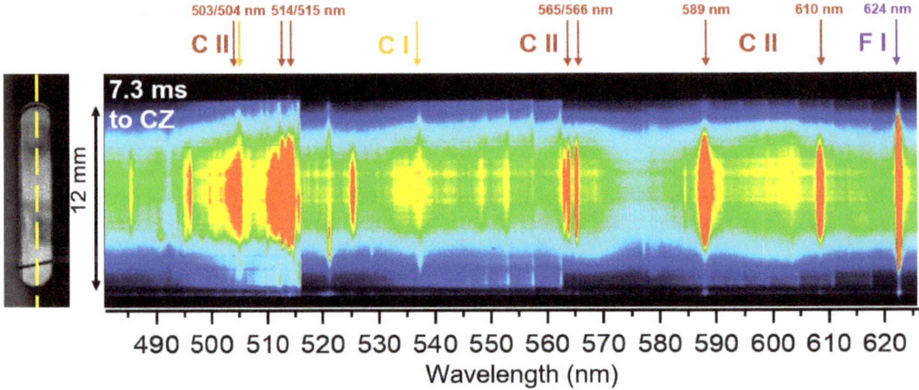

Figure 3. (**left**) Photo (grey) of observation window. (**right**) Two-dimensional optical emission spectroscopy frame at 7.3 ms before current zero (CZ).

Secondly, other Swan band pattern ws observed over the full vertical axis of the side-on 2D spectra. The example shown in Figure 4 was acquired with setup (b) about 6 ms before CZ, i.e., shortly before the peak current. A grating of 1800 L/mm was applied to obtain higher spectral resolution. A good agreement was found of the 1D-spectrum taken in central position with spectra shown by Camacho in OES investigations on plumes produced by laser ablation of graphite targets [18]. The weaker continuum and stronger C II lines compared to [18] hint on rather high plasma temperatures at least in the arc center with higher current density than near to the wall.

Exemplarily, one of the lines of the C_2 Swan band near 562.8 nm was analyzed; the carbon ionic line at 566.2 nm was used for comparison, cf. yellow arrows in the 2D spectrum. The side-on radiances are shown in the lower-left part of Figure 4: whereas the ionic line has its maximum in the center, the Swan band emission is spread more homogeneously over a wide side-on positions between center and 4 mm but has a distinct maximum near to 5 mm, i.e., near to the wall. Since both emissions showed good symmetry in relation to the center, this axis was used for symmetrization and as the central side-on position "0 mm". Then, the radial profile of the emission in the arc can be analyzed by Abel inversion of the side-on radiances. Results are shown in the lower right part of Figure 4. The C II 566.2 nm ionic line is emitted as expected mainly in the center; the emission coefficient decreases to 20% within radial positions of 2 mm. The C_2 Swan band, however, has a sharp peak of less than 2 mm FWHM with a maximum emission coefficient below 1 mm to the wall. It should be noted that although the nozzle diameter is 12 mm some intensity was detected at side-on position above 6 mm due to experimental limitations like quartz plate connection and refraction at the windows. The algorithm of inverse Abel transformation is limited in case of very low emission from the central position, therefore the C_2 emission coefficient in the center is not plotted for values below 10% (radial positions < 3.5 mm). To summarize even in the case of Figure 4 the Swan bands are emitted only in a thin sheath at the wall.

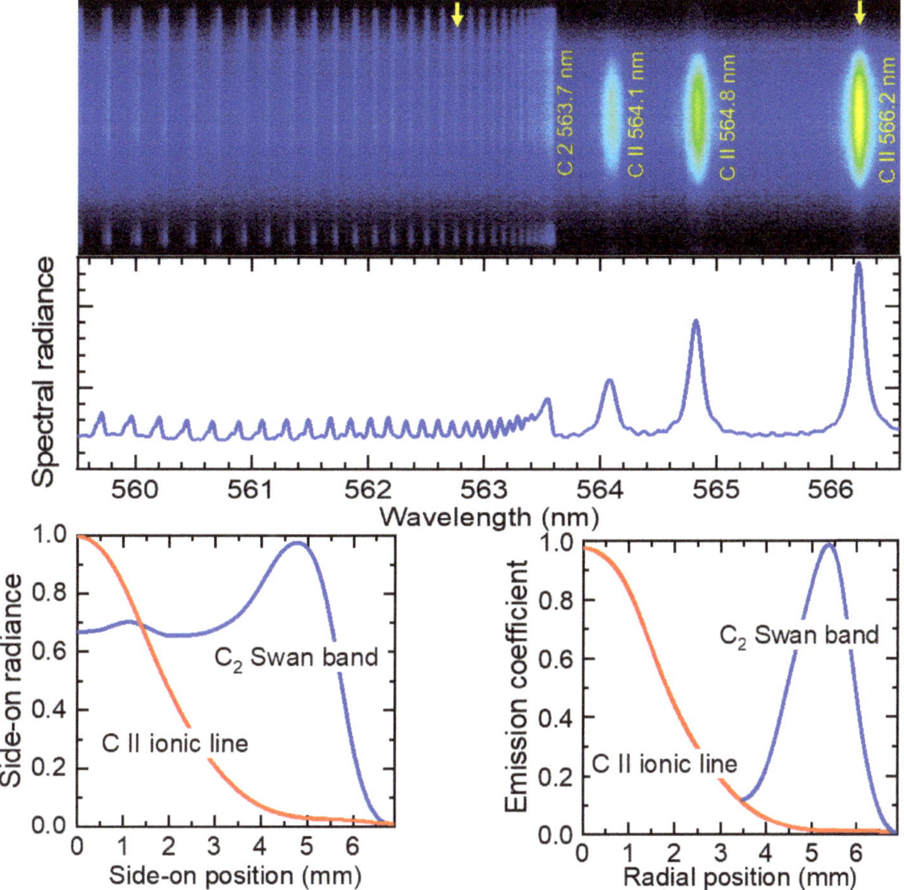

Figure 4. (**Top**): 2D spectrum over the full arc cross-section (upper part) together with the corresponding 1D spectrum from central arc position (lower part) in the spectral range around the C_2 Swan band head at 563.7 nm-Swan band on the left side and C II lines on the right side. (**Bottom left**): Spectrally integrated line intensities of the carbon ion line C II 566.2 nm and of the C_2 Swan band line at 562.8 nm, labeled by yellow arrows. (**Bottom right**): Inverse Abel transformation carried out for these intensities to reveal the origin of emission.

A third example of the occurrence of C_2 Swan bands is shown in Figure 5. It was only observed with setup (a) providing higher pressure and strong wall ablation due to peak currents of 8 kA (100 Hz). With the single long PTFE nozzle the current was not switched off and multiple current zero transitions were observed. Except the ignition phase and few hundred µs around CZ, all spectra are dominated by pronounced emission of the Swan bands. The band heads of the Swan bands are located at 473.7 nm, 516.5 nm, and 563.6 nm; they are indicated by red arrows in the two-dimensional spectrum in the upper part of Figure 5. The wavelength range chosen here does not include the band head at 438.2 nm but also contains the C II lines at 564.06 nm, 564.81 nm, and 566.26 nm as well as the C I atomic lines at 476.2 nm, 477.0 nm, 493.2 nm, 505.2 nm, and 538.0 nm. Weak or non-visible ionic and atomic carbon lines in comparison with the Swan bands give the first hint to rather low temperatures in the center of the arc. Furthermore, it was observed that the occurrence of carbon lines drastically changes approaching current zero. Within some 100 μs, first the ionic and then the atomic lines disappear; after CZ they reappear in reversed order. In fact, disappearance of the atomic lines cannot be observed for first and second, but for the third CZ crossing.

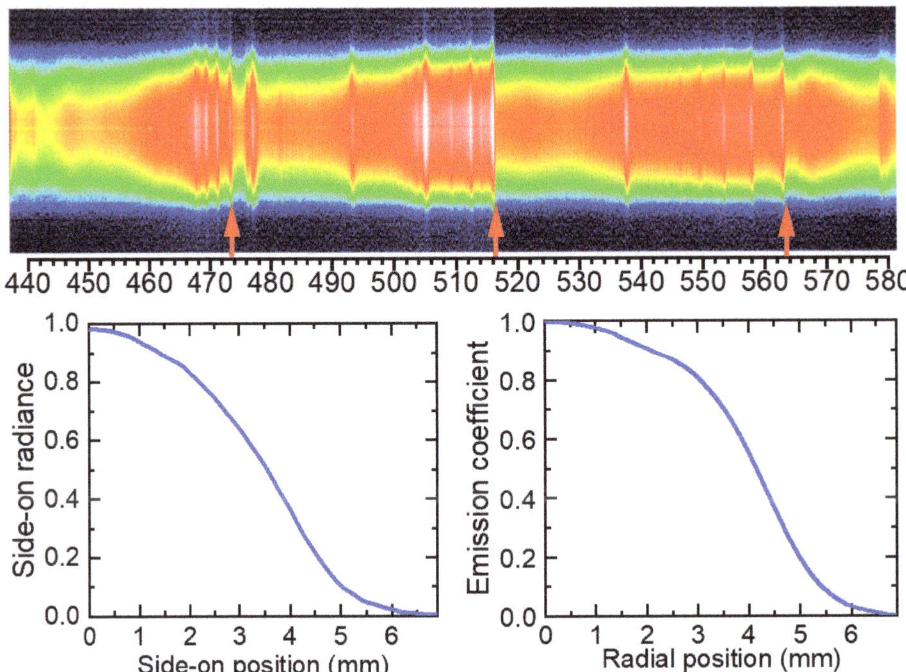

Figure 5. (**top**) Spectrum acquired 300 μs before CZ with setup (a) and 8 kA peak current. It is completely dominated by molecular radiation of C_2 Swan bands (band heads labeled by arrows). (**bottom**) Spectrally integrated line intensities (**left**) and emission coefficient obtained from Abel inversion (**right**) of the C_2 Swan band emission at 562 nm.

The Swan band pattern has a much higher intensity in the central position, although the emission is extended to the side-on positions of the nozzle wall. The origin of emission is further analyzed using the band head around 563 nm as shown in the lower part of Figure 5. The side-on profile (left) and the emission coefficient obtained by inverse Abel transformation (right) reveal a different occurrence in comparison to the plasma in Figure 4. The Swan bands were emitted with the highest intensities in the center of the arc, continuously decreasing towards the nozzle walls. Thus, it can follow that the arc plasma is completely dominated by the PTFE material and it is characterized by rather low temperatures even in the arc center. It should be mentioned that this third case of Swan band appearance is the most extreme and could not be achieved with setup (b) with two nozzles separated by the heating channel even when the peak current was doubled to 10 kA.

The fourth example was typical for setup (b): Here, the Swan bands can be recognized by their characteristic absorption pattern at currents >4 kA, cf. example in Figure 6. The spectrum was taken shortly after the current maximum (4.7 kA, 4.6 ms before CZ). Emission from the hot plasma in the arc center served as an internal background radiator that was absorbed by the much cooler carbon dimers near to the nozzle wall. As in Figure 5, the Swan band heads are indicated by red arrows. Additionally to the absorption pattern, some emission lines can be found. These are all ionic carbon lines, e.g., at C II 564.06 nm, 564.81 nm, and 566.26 nm. They are preferably emitted in the arc center, i.e., at higher temperatures. In the third and fourth cases with most intense ablation and material transport towards the electrodes, which is probably the reason why no copper lines were observed at the slit position. Fluorine lines were not available in that spectral range.

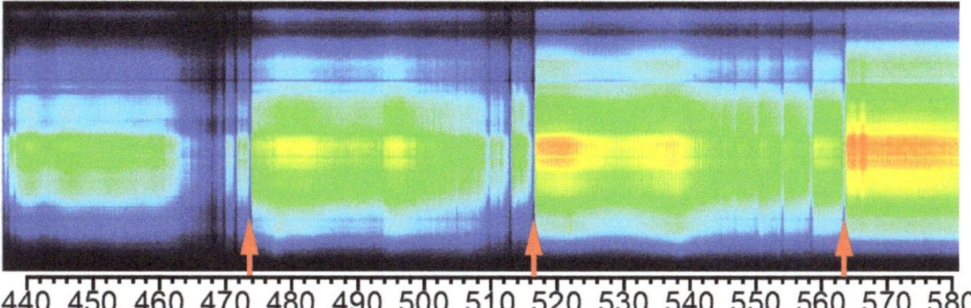

Figure 6. Spectrum acquired with setup (b) during high-current phase. The typical structure of the C_2 Swan bands was found as an absorption pattern with the plasma in the arc center serving as background radiator. The band heads are labeled by arrows.

3.2. Optical Absorption Spectroscopy around Current Zero

The phase of current zero-crossing is of the highest importance for an understanding of the switch-off process and the dielectric recovery of the electrode gap region. Hence it is of special interest to extend experimental knowledge as close as possible to CZ and even beyond. However, even tapping the full potential of optical emission spectroscopy, e.g., by application of OES with intensified cameras as described above, the analysis based on optical emission spectroscopy is limited to times about 10 µs before CZ due to reduced energy input by the arc [9]. Consequently, absorption techniques were required for further investigation of the current zero-crossing and the immediately following time period. Since the majority of atoms are in the ground state in case of the lower temperatures near CZ, it will be necessary to mainly analyze lines going to ground or very low levels by optical absorption spectroscopy (OAS). However, most of the relevant lines are in deep UV regions far below 300 nm. From the experimental point it is extremely demanding to investigate such radiation under switching-relevant conditions since all components of the setup including high-pressure vessel and model circuit chamber have to be transparent for these wavelengths. With the actual setup even resonant lines that might be more suitable could not be detected due to limited spectral sensitivity of the cameras such as C I at 296 nm or Cu I at 324 nm and 327 nm. The few resonant lines in the available wavelength range above 340 nm have very low transition probabilities, e.g., C I at 462 nm and O I at 630 nm. However, it might be possible that some lines might be occupied around CZ and could be detected by OAS that are characterized by relatively low energy levels and medium transition probabilities, e.g., Cu I at 510 nm, 570 nm, and 578 nm with E_u = 1.39 eV and 1.64 eV or O I at 557 nm with E_u = 1.26 eV. Additionally, molecules are possible candidates for absorption, e.g., the C_2 molecule since its Swan bands were observed in emission until few 100 µs before CZ and even in absorption during the high-current phase as shown above.

Broadband optical absorption spectroscopy (OAS) was carried out around CZ using the pulsed high-intensity xenon lamp as an external wide-band background illumination. Two examples are shown in Figure 7, comprising the wavelength ranges about 440–600 nm (left) and 640–800 nm (right column). In the upper panel only the emission from the Xe lamp is given, i.e., through model circuit breaker including all windows but without discharge. Broadband continuum can be seen in both spectral ranges. It should be noted that the spectra are not calibrated concerning absolute intensity. The edges of the nozzle can be recognized by sharp transitions from the bright stripe (white/red) from the Xe lamp illumination and the dark regions (green). The spatial distribution within the nozzle slit is quite homogeneous, showing smooth illumination by the background source. In the middle panel, the OES spectra were taken at CZ (exposure time 50 μs) with the Xe lamp continuum passing the remainder of arc discharge in the nozzle. Patterns of horizontal stripes were sometimes observed. Similar experiments using HSC instead of ICCD camera revealed that these stripes did not change from one video frame to the next. Thus, it was reasonable to assume a deposition on the quartz glass sealing the slits, e.g., by particles. In the left spectrum a certain structure was found below 500 nm whereas the right spectrum did not show any peculiarities. From spectra in top and middle panel a transmission could be calculated, cf. lower panel of Figure 7. For an improvement of the signal-to-noise ratio and better visualization of the intensity ratio, spatial integration was carried out for the determination of the transmission. It revealed that there was only one significant absorption peak around 493 nm. This absorption was clearly accorded to the CuF molecule as will be discussed below. Beside this CuF peak, no hint on any absorbing lines or other features could be detected around current zero, even with the intensified camera with high sensitivity and dynamic range. Even the C_2 Swan bands could not be observed before or after CZ in OAS with the Xe lamp as background radiator although they were detectable in OES up to a few hundred μs before CZ. Moreover, a closer look onto the emission spectra (cf. Figure 3) showed that the CuF absorption at 493 nm could also be found during the high-current phase of the discharge, though this effect was rather weak compared to the intense line emission. This will be further investigated in Section 3.3.

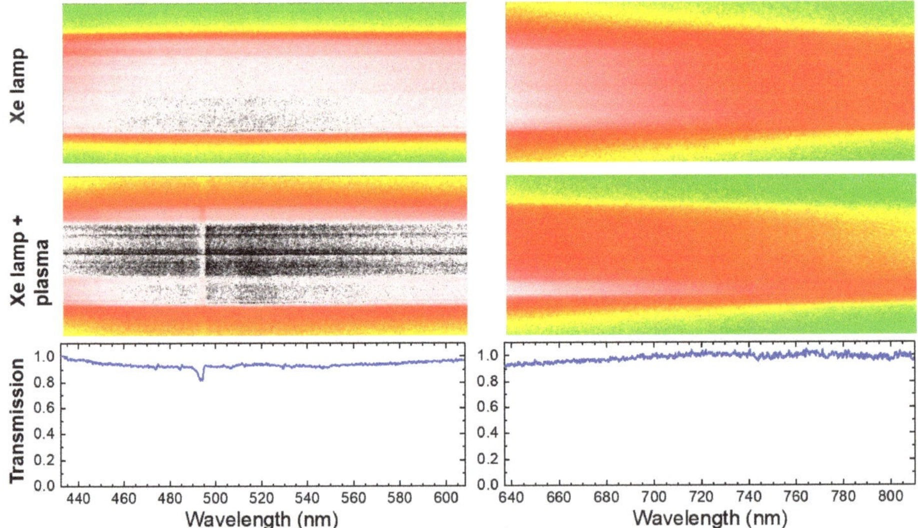

Figure 7. Optical absorption spectroscopy. (**Top**): Overview emission spectra of Xe lamp only. (**Middle**): Xe lamp with discharge of 5 kA peak current and setup (b), acquired at CZ with 50 μs exposure time. (**Bottom**): Spatially integrated transmission calculated from above spectra showing absorption around 500 nm.

Other species for absorption with maximum around 493 nm could be excluded in detailed spectral analysis, including all relevant elements as Cu and W from electrodes, C, O, and F from filling gas and nozzle, and even H as possible contamination. As an example, a prominent candidate might have been the carbon atomic line C I at 493.20 nm, although its lower energy level of 7.7 eV is rather high. However, this line was not detected in emission like other atomic carbon lines with similar upper level of about 10 eV and comparable transition probabilities in the range of several 10^6 s^{-1}, e.g., C I 505.21 nm and 538.03 nm (cf. Figure 3). Moreover, these C I lines were still observed in emission 0.7 ms before CZ, while at 493 nm an absorption could be seen even during discharge.

No absorption spectra were found in the literature for the CuF molecules. Thus, in Figure 8, an emission spectrum from Cheon et al. [25] (black curve) was added to the calculated absorption spectrum (dashed blue) for comparison. Considering different experimental conditions and methods, a compelling agreement was found (OES with higher spectral resolution will be shown below). Basic data of the CuF emission are listed in Table 1.

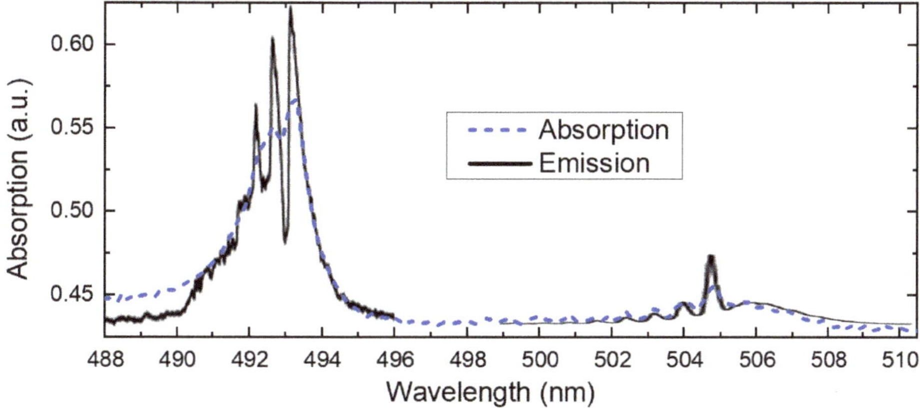

Figure 8. Spectral absorption measured around CZ (dashed blue) compared with CuF emission from [25] (black line).

Table 1. Basic data of CuF emission lines [26,27].

Wavelength nm	Relative Intensity	Lower Level eV	Upper Level eV	Transition		Quantum Upper	Number Lower	System
478.19	400	0	2.51	$X^1\Sigma^+$	$C^1\Pi$	0	1	C
490.13	500	0	2.44	$X^1\Sigma^+$	$B^1\Sigma$	0	1	B
492.68	600	0	2.51	$X^1\Sigma^+$	$C^1\Pi$	1	1	C
493.20	800	0	2.51	$X^1\Sigma^+$	$C^1\Pi$	0	0	C
505.23	600	0	2.44	$X^1\Sigma^+$	$B^1\Sigma$	1	1	B
506.11	700	0	2.44	$X^1\Sigma^+$	$B^1\Sigma$	0	0	B
508.64	200	0	2.51	$X^1\Sigma^+$	$C^1\Pi$	1	0	C
567.72	500	0	2.18	$X^1\Sigma^+$	$A^1\Pi$	2	2	A
568.57	600	0	2.18	$X^1\Sigma^+$	$A^1\Pi$	1	1	A
569.43	600	0	2.18	$X^1\Sigma^+$	$A^1\Pi$	0	0	A

In the following, the CuF molecular absorption after current zero should be analyzed in more detail. A series of time-resolved spectra is shown in the upper part of Figure 9. The transmission was calculated based on the division of the measured spectra (plasma plus xenon lamp) by a xenon lamp spectrum without discharge. A higher spectral resolution was obtained by the grating with 1800 L/mm. Thus, the peak structure including maxima at 493.2 nm, 492.7 nm, and 493.2 nm can be clearly recognized in agreement with the emission spectrum of CuF molecules from [25] shown in Figure 8. The overlaying periodic structure is not caused by the plasma in the nozzle since the same structure was also observed for the xenon lamp itself. Probably it was caused by an interference effect of glass plates in the detector. The background intensity increases although the xenon lamp is in its plateau phase.

For a quantization of the temporal evolution of the absorption, the area under the curve (AUC) was determined from the difference between the "background" (average of levels extracted at wavelengths aside the CuF absorption, i.e., around 490 nm and 496 nm) and the transmission, integrated over the spectral range. The AUC is plotted as the curve with black open circles in the lower part of Figure 9. Additionally, a normalization of the AUC was carried out by division by the (temporally increasing) intensity of the background signal. The normalized AUC is represented by the curve with red filled squares. The spectrum 0.2 ms after CZ was chosen as the starting point and the according value was set to 1 for better comparison. Within half a millisecond, the AUC decreases by 60%. The decrease of the normalized AUC is even more significant, namely down to 20% (factor of 5). Few other shots that were carried out confirmed this result. However, due to the exponential nature of this decrease, the absolute values are sensitive to the starting point. Summarizing it can be stated that the CuF absorption and thus, also the CuF density decreases after current zero on a timescale of several hundreds of microseconds.

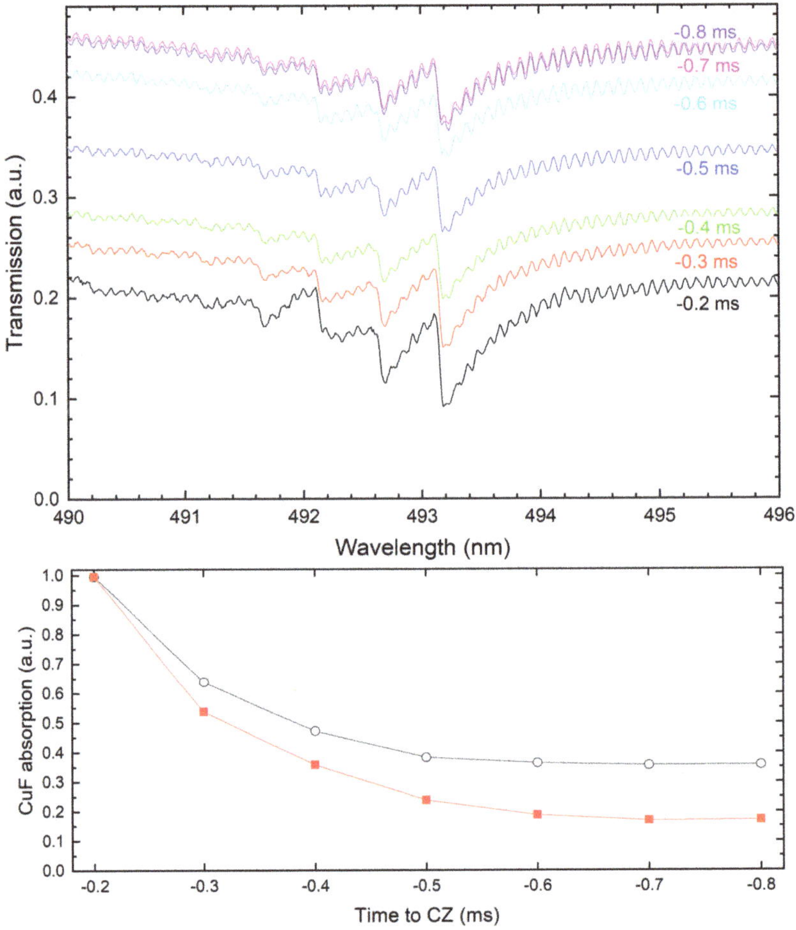

Figure 9. (**top**) Series of detail transmission spectra after current zero. (**bottom**) Temporal development of absorption peak at 493 nm calculated as area under curve (AUC, open circles) and AUC normalized by background intensity (filled squares).

3.3. CuF during the High-Current Phase

As mentioned above, overview OES spectra in Figure 3 gave hint on a possible absorption of the CuF molecule even during the arc discharge though the effect might be considerably lower in absolute intensity than the atomic line emission. Thus, the spectral range around the 493 nm-peak was investigated with video OES of higher resolution (grating 1800 L/mm instead of 150 L/mm, exposure time 98 μs). An example is shown in Figure 10 acquired with setup (b) and 5.3 kA peak current.

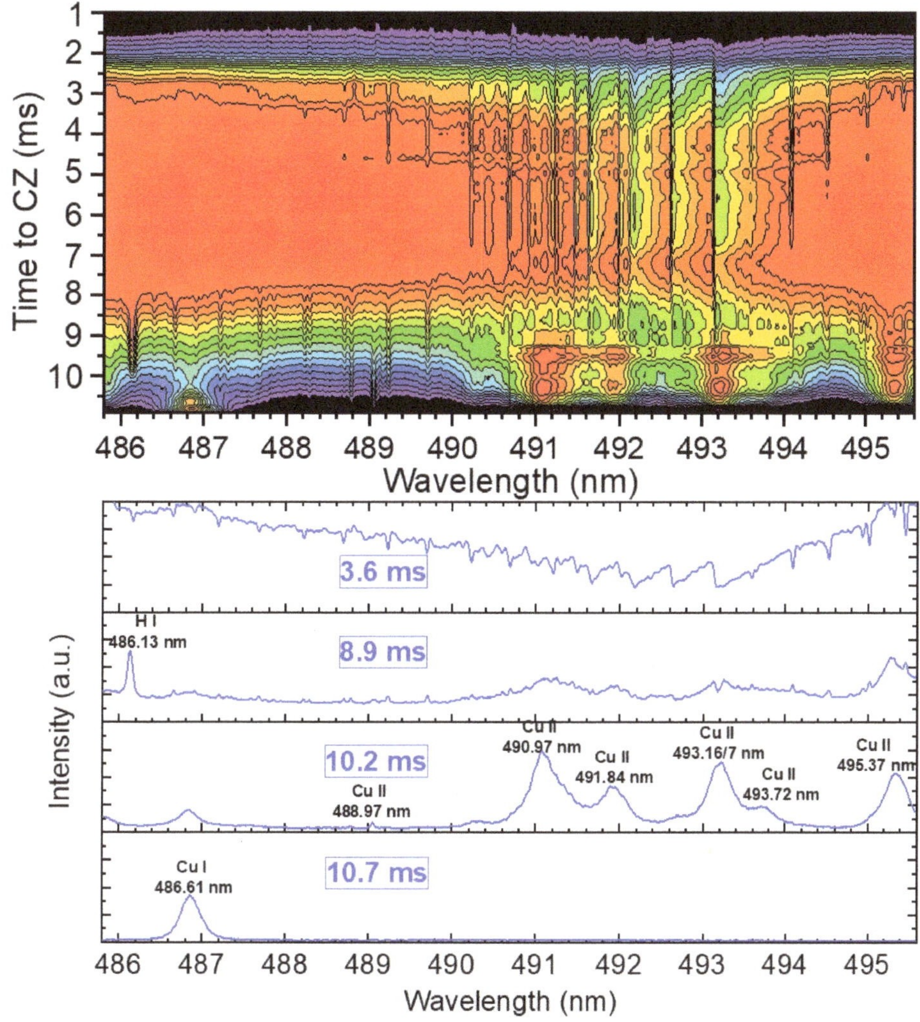

Figure 10. Temporal evolution of the spectral range around 493 nm with CuF absorption (Time to CZ is from (**bottom**) to (**top**), the color scale reaches from black for lowest emission to blue, green, and red for highest intensities).

In the upper part, all about 110 optical emission spectra (spatially integrated) are plotted line by line vs. time to CZ, forming a two-dimensional contour plot. In the lower part of Figure 10 a selection of four instants of time with characteristic spectral features are shown. Additionally, several atomic and ionic lines were labeled that helped for fine adjustment and control of exact wavelength positions. The ignition phase the spectra were dominated by atomic and ionic copper line emission. At first, i.e., 10.7 ms before current zero, an atomic copper line Cu I 486.61 nm was observed which was followed by several ionic copper lines (cf. spectrum 10.2 ms before CZ). As known from overview OES spectra, the ablation of PTFE was usually initialized about 1 ms after ignition. In the spectral range of Figure 10 no atomic fluorine lines can be observed. However, the occurrence of the hydrogen line H_β at 486.13 nm (starting about 9 ms before CZ) can be regarded as an early sign of nozzle ablation, probably caused by a thin remaining water film on the nozzle surface. Thus, one may expect the occurrence of CH molecules. That should be detectable by the CH(A-X) band with maxima around 430 nm. However, we did not find such an absorption pattern even in a detailed analysis of the corresponding spectral range. Within several hundred microseconds the spectrum is changed from being dominated by copper lines (more probable originating from the W–Cu electrodes than from the ignition wire) to being ablation-dominated. CuF absorption pattern is observed during the full high-current phase, at least from about 8 to 2 ms before CZ. In the spectral range below 500 nm this is basically visible by a broad continuum, starting about 2–3 ms after ignition (or 8 ms before CZ) and lasting at least until about 2 ms before CZ. As can be clearly seen from the top spectrum in Figure 10 the characteristic absorption pattern of CuF can be observed for these 5–6 ms, i.e., during the whole high-current phase of the discharge. Comparable to the case of absorption of the C_2 Swan bands, the CuF absorption is enabled by background continuum from the arc plasma. The observed temporal fluctuation of the intensity has been found to be caused mainly by changes of background intensity, e.g., fluctuation of transmission or reflection due to droplets. Hence, even during the high-current phase a considerable amount of absorbing CuF molecules must be existent in the plasma at nozzle position, i.e., 8 mm away from the electrode. This might be unexpected but leads to the conclusion that the gas flow into the heating channel is strong enough to pull electrode material into the region of optical investigation.

4. Discussion

Information was obtained from spatially and temporally resolved video spectroscopy using HSC. That comprised the different phases of discharge and the occurrence of Swan band emission from C_2 molecules. These Swan bands could be observed under varying conditions. Different amount of ablated PTFE from the nozzle wall and plasma temperature were generated depending on nozzle geometry and current density. According to the equilibrium composition calculations by Yang et al. [11], a considerable radiation of the C_2 molecule indicates plasma temperatures in the range from 4000–6000 K. Firstly, there was an occurrence very close to the nozzle walls as typical behavior for cases of moderate PTFE influence, i.e., when the current density was not too high and the temperature close to the wall was rather low, allowing the existence of carbon dimers. Although it might be often neglected when the temperature distribution in the arc is investigated, the Swan bands represent the existence of carbon molecules due to wall ablation and thus, an important effect of cooling and change of plasma composition. Secondly, with higher current densities, the Swan band patterns were also distributed over the full vertical axis of the side-on 2D-spectra. However, it was found by Abel inversion that the Swan bands are emitted in a thin sheath at the nozzle wall. Thirdly, a different distribution was found under extreme conditions, i.e., with single long PTFE nozzle and high peak currents. The arc plasma was completely dominated by PTFE material and temperatures were moderate in the arc center, proved by weak or non-visible ionic and atomic carbon line emission. The Swan band pattern was emitted with the highest intensity in the central position though emission was extended to radial positions of the nozzle wall. Finally, Swan bands also appeared as an absorption pattern at moderate currents with setup (b). Emission from hot plasma in the arc center (proved by C II line emission) served as an internal background radiator that was absorbed by the much cooler carbon dimers near to the nozzle wall.

A considerable amount of CuF molecules in the high-current arc as well as around CZ was found from absorption spectra. This was not expected before, for several reasons, especially regarding that no other molecules were observed close to current-zero. A possible explanation is as follows: Around CZ it is expected that convective fluxes are significantly reduced due to equalization of pressures. As a consequence, copper atoms from still hot electrodes may expand diffusively along the nozzle and reach the position of OES (nozzle slit). In parallel, fluorine-containing molecules are still released from the nozzle wall. CuF molecules could be formed by chemical reaction of atomic F and Cu either at the hot W–Cu electrode surface followed by CuF evaporation or in the gas phase with Cu evaporated from the electrode. Similarly, the observed absorption during the high-current phase might be explained by the gas flow out of the nozzle into the heating chamber. In this case, copper atoms eroded or evaporated from the W–Cu electrode might be flushed with the stream towards the heating channel, reacting on its way with fluorine from the wall, and being detected at the observation slit by absorption with the arc plasma as background radiator. However, during the time immediately after flow reversal, i.e., about 1 ms before CZ, the situation is very different: the gas flow is directed from the heating channel towards the electrodes. Thus, no copper from the electrodes should reach the observation area with the slit and react with fluorine. That means that probably no CuF should be produced at this period; any detected CuF should be a survivor from the heating chamber. As a pity, at the moment database is not sufficient to answer the question if there is a lower CuF concentration after flow reversal or not. In the video spectra there is simply not enough background emission to enable sufficient signal for an absorption.

Within the described experiments, limitations of reproducibility, fluctuation in transmission due to particles, film layers on windows, and dust did not allow temporally and spatially resolved determination of the absorption by CuF, e.g., using two-dimensional inverse Abel transformation of video spectra with higher spectral resolution. Nevertheless, this would be the next step if significant technical improvements were done. On the one hand, further optimization of the nozzle slit, its position and manufacturing technique might provide even fewer changes of the gas and droplet flow conditions, thus allowing measurements still closer to the undisturbed conditions at the nozzle. On the other hand, the observation technique itself might be improved, too. Nowadays, the advantages of intensified and high-speed video cameras can be combined in new generations of cameras or boosters. The background illumination could be improved, too. Beside improvements in the optical path in order to enhance the intensity and homogeneity, the pulsed xenon lamp might also be replaced by a laser-driven light source with extended pulse duration. As a consequence, quantification of the CuF absorption after CZ as well as during the arc discharge might be possible. Furthermore, OAS regarding Swan bands could be tackled. Last not least, tests with other electrodes should be carried out to finally prove the origin of absorption by CuF-molecules, e.g., made of pure tungsten.

Altogether, the possibilities of a recording of molecule radiation emission and absorption in the visible spectral range have been demonstrated for the case of high-current ablation dominated arcs. Using PTFE nozzles, tungsten–copper electrodes and operation in air or CO_2, the Swan bands of the carbon dimer C_2 and absorption of the CuF molecule were the only detectable radiation patterns. However, these patterns open up ways for a study of interesting ranges in high-current breaking processes like the colder plasma ranges near the nozzle walls and the time around CZ.

5. Conclusions

Ablation-dominated switching arcs have been investigated in a model circuit breaker with CO_2 atmosphere as well as in a long PTFE nozzle under ambient conditions. Optical emission spectroscopy and broadband optical absorption spectroscopy were carried out using either intensified or high-speed cameras. As a main result, we have shown that specific molecules are detectable in the wavelength range between 400 nm and 800 nm under strongly varying conditions. It was demonstrated that depending on nozzle geometry and discharge current the C_2 Swan bands can be observed by their emission (i) near to the wall only, (ii) distributed over the full arc diameter with the highest intensity in the center, or (iii) by their absorption of continuum radiation from the arc plasma. Although an emission was found until a few hundreds of microseconds before CZ, no absorption of the Swan bands could be detected around and after CZ. Even the occurrence of C_2 radiation can be used an indicator for intermediate temperatures of around 5000 K according to composition calculations. An accurate determination of quantities like rotational temperatures and densities can be performed by comparison with spectra calculation. This was out of the scope of the present paper and could be tackled in the future, demanding two-dimensional inverse Abel transformation of multiband spectra. Additionally, these findings could be applied e.g., for the study of fluxes and distribution of evaporated material and for the verification of erosion models of the PTFE wall. The molecule CuF could be expected when copper vapor from the electrode erosion is mixed with the dissociated PTFE vapor from the nozzle ablation. To our knowledge, it was the first observation of CuF molecules in high-current arcs burning in nozzles under ablation-dominated regime. The agreement of spectral features between the literature on CuF and our experiments seem very plausible. However, further investigations are necessary to confirm these findings. The CuF molecular absorption could be applied as an alternative to investigating electrode erosion and distribution of electrode material within the discharge area, especially around CZ when the emission of atomic copper lines fades out.

Author Contributions: Conceptualization, R.M. and D.U.; methodology, validation, and formal analysis, R.M.; investigation with setup (a), R.M.; investigation with setup (b), R.M. and N.G.; writing—original draft preparation, R.M. and D.U.; project administration, D.U. All authors have read and agreed to the published version of the manuscript.

Funding: This research was funded by Deutsche Forschungsgemeinschaft grant numbers UH 106/13-1 and SCHN 728/16-1.

Acknowledgments: The authors would like to thank Steffen Franke and Alireza Khakpour for experimental help and fruitful discussions. The calculation of plasma composition was realised by Sergey Gortschakow (all Leibniz Institute for Plasma Science and Technology).

Conflicts of Interest: The authors declare no conflict of interest. The funders had no role in the design of the study; in the collection, analyses, or interpretation of data; in the writing of the manuscript, or in the decision to publish the results.

Abbreviations

The following abbreviations are used in this manuscript:

AUC	area under curve
CO_2	carbon dioxide
CuF	copper fluoride
CZ	current zero
ICCD	intensified charge coupled device
HSC	high-speed camera
MoS_2	molybdenum disulfide
OAS	optical absorption spectroscopy
OES	optical emission spectroscopy
PTFE	polytetrafluoroethylene
SF_6	sulfur hexa–fluoride
W–Cu	tungste–copper

References

1. Ruchti, C.B.; Niemeyer, L. Ablation controlled arcs. *IEEE Trans. Plasma Sci.* **1986**, *14*, 423–434. [CrossRef]
2. Seeger, M.; Tepper, J.; Christen, T.; Abrahamson, J. Experimental study on PTFE ablation in high voltage circuit-breakers. *J. Phys. D Appl. Phys.* **2006**, *39*, 5016–5024. [CrossRef]
3. Eichhoff, D.; Kurz, A.; Kozakov, R.; Gött, G.; Uhrlandt, D.; Schnettler, A. Study of an ablation–dominated arc in a model circuit chamber. *J. Phys. D Appl. Phys.* **2012**, *45*, 305204. [CrossRef]
4. Franke, S.; Methling, R.; Uhrlandt, D.; Gorchakov, S.; Reichert, F.; Petchanka, A. Arc temperatures in a circuit breaker experiment from iterative analysis of emission spectra. *J. Phys. D Appl. Phys.* **2020**, *53*, 385204. [CrossRef]
5. Bort, L.; Schultz, T.; Franck, C.F. Determining the time constant of arcs at arbitrary current levels. *Plasma Phys. Technol.* **2019**, *5*, 175–179. [CrossRef]
6. Tanaka, Y.; Yokomizu, Y.; Matsumura, T.; Kito, Y. The opening process of thermal plasma contacts in a post-arc channel after current zero in a flat-type SF6 gas-blast quenching chamber. *J. Phys. D Appl. Phys.* **1997**, *30*, 407–416. [CrossRef]
7. Hartinger, K.T.; Pierre, L.; Cahen, C. Combination of emission spectroscopy and fast imagery to characterize high-voltage SF6 circuit breakers. *J. Phys. D Appl. Phys.* **1998**, *31*, 2566–2576. [CrossRef]
8. Kozakov, R.; Kettlitz, M.; Weltmann, K.D.; Steffens, A.; Franck, C.M. Temperature profiles of an ablation controlled arc in PTFE: I. Spectroscopic measurements. *J. Phys. D Appl. Phys.* **2007**, *40*, 2499–2506. [CrossRef]

9. Methling, R.; Khakpour, A.; Götte, N.; Uhrlandt, D. Ablation-Dominated Arcs in CO_2 atmosphere—Part I: Temperature Determination near Current Zero. *Preprints* **2020**, 2020080279. [CrossRef]
10. Paul, K.C.; Sakutay, T.; Takashimay, T.; Ishikawa, M. The dynamic behaviour of wall-stabilized SF_6 arcs contaminated by Cu and PTFE vapours. *J. Phys. D Appl. Phys.* **1997**, *30*, 103–112. [CrossRef]
11. Yang, A.; Liu, Y.; Sun, B.; Wang, X.; Cressault, Y.; Zhong, L.; Rong, M.; Wu, Y.; Niu, C. Thermodynamic properties and transport coefficients of high-temperature CO_2 thermal plasmas mixed with C_2F_4. *J. Phys. D Appl. Phys.* **2015**, *48*, 495202. [CrossRef]
12. Khalid, R.; Yaqub, K.; Yaseen, S.; Javeed, S.; Ashraf, A.; Janjua, S.A.; Ahmad, S. Sputtering of graphite in pulsed and continuous arc and spark discharges. *Nucl. Instrum. Meth. Phys. Res. B* **2007**, *263*, 497–502. [CrossRef]
13. Bystrzejewski, M.; Łabędź, O.; Lange, H. Diagnostics of carbon arc plasma under formation of carbon-encapsulated iron nanoparticles by optical emission and absorption spectroscopy. *J. Phys. D Appl. Phys.* **2013**, *46*, 355501. [CrossRef]
14. Vekselman, V.; Feurer, M.; Huang, T.; Stratton, B.; Raitses, Y. Complex structure of the carbon arc discharge for synthesis of nanotubes. *Plasma Sources Sci. Technol.* **2017**, *26*, 065019. [CrossRef]
15. Becerra, M.; Friberg, A. Arc jets blown by outgassing polymers in air. In Proceedings of the 20th International Conference on Gas Discharges and Their Applications GD2014, Orleans, France, 6–11 July 2014; Volume 1, pp. 1–4.
16. Harilal, S.S.; Issac, R.C.; Bindhu, C.V.; Nampoori, V.P.N.; Vallabhan, C.P.G. Optical emission studies of C_2 species in laser–produced plasma from carbon. *J. Phys. D Appl. Phys.* **1997**, *30*, 1703–1709. [CrossRef]
17. Park, H.S.; Nam, S.H.; Park, S.M. Optical emission studies of a plume produced by laser ablation of a graphite target in a nitrogen atmosphere. *Bull. Korean Chem. Soc.* **2004**, *25*, 620–626.
18. Camacho, J.J.; Diaz, L.; Santos, M.; Reyman, D.; Poyato, J.M.L. Optical emission spectroscopic study of plasma plumes generated by IR CO_2 pulsed laser on carbon targets. *J. Phys. D Appl. Phys.* **2008**, *41*, 105201. [CrossRef]
19. Parigger, C.G.; Hornkohl, J.O.; Nemes, L. Time-resolved spectroscopy diagnostic of laser-induced optical breakdown. *Int. J. Spectrosc.* **2010**, *2010*, 593820. [CrossRef]
20. Witte, M.J. Diatomic Carbon Measurements with Laser-Induced Breakdown Spectroscopy. Master's Thesis, University of Tennessee, Knoxville, TN, USA, 2015.
21. Reynaud, O.; Picard, J.P.; Parizet, M.J. Rotational Temperatures Determined from C_2 Molecular Band Spectra in a Thermal Argon Plasma Interacting with Insulating Materials. *Spectrosc. Lett.* **1995**, *28*, 1007–1014. [CrossRef]
22. Sakuyama, T.; Amitani, K.; Tanaka, Y.; Uesugi, Y.; Kaneko, S.; Okabe, S. Investigation on Spatial Temperature Decay of Thermal Plasma with Polymer Ablation. *J. Plasma Fusion Res. Ser.* **2008**, *8*, 732–735.
23. Hong, D.; Sandolache, G.; Lan, K.; Bauchire, J.M.; Le Menn, E.; Fleurier, C. A radiation source developed for broad band optical absorption spectroscopy measurements. *Plasma Sources Sci. Technol.* **2003**, *12*, 1. [CrossRef]
24. Günther, K.; Radtke, R. A proposed radiation standard for the visible and UV region. *J. Phys. E Sci. Instrum.* **1975**, *8*, 371–376. [CrossRef]
25. Cheon, J.; Kang, H.K.; Zink, J.I. Spectroscopic identification of gas phase photofragments from coordination compound chemical vapor deposition precursors. *Coord. Chem. Rev.* **2000**, *200–202*, 1009–1032. [CrossRef]
26. Pearse, R.W.B.; Gaydon, A.G. *The Identification of Molecular Spectra*, 3rd ed.; Chapman and Hall: London, UK, 1963.
27. SpecLine-Spectral Line Idenfication for Atoms and Molecules. Available online: https://www.plasus.de (accessed on 10 September 2020).

Article

Effects of Fast Elongation on Switching Arcs Characteristics in Fast Air Switches

Ali Kadivar [1,2,*] and Kaveh Niayesh [1]

[1] Department of Electric Power Engineering, Norwegian University of Science and Technology, 7034 Trondheim, Norway; kaveh.niayesh@ntnu.no

[2] Department of Transmission Line and Substation Equipment, Niroo Research Institute (NRI), End of Dadman Street, Shahrak Ghods, Tehran 1468613113, Iran

* Correspondence: ali.kadivar@ntnu.no

Received: 25 July 2020; Accepted: 14 September 2020; Published: 16 September 2020

Abstract: This paper is devoted to investigating the effects of high-speed elongation of arcs inside ultra-fast switches ($u_{contact} \approx$ 5–80 m/s), through a 2-D time-dependent model, in Cartesian coordinates. Two air arcs in series, one between a stationary anode and a moving cathode and the other between a stationary cathode and a moving anode in the arc chamber, are considered. A variable speed experimental setup through a Thomson drive actuator is designed to support this study. A computational fluid dynamics (CFD) equations system is solved for fluid velocity, pressure, temperature, and electric potential, as well as the magnetic vector potential. Electron emission mechanisms on the contact surface and induced current density due to magnetic field changes are also considered to describe the arc root formation, arc bending, lengthening, and calculating the arc current density, as well as the contact temperatures, in a better way. Data processing techniques are utilized to derive instantaneous core shape and profiles of the arc to investigate thermo-electrical characteristics during the elongation progress. The results are compared with another experimentally verified magnetohydrodynamics model of a fixed-length, free-burning arc in the air. The simulation and experimental results confirm each other.

Keywords: air arc plasma; Thomson actuator; magnetohydrodynamic simulations; fast switch

---

1. Introduction

Studies show that about 20% of industries have to change 5% to 10% of their circuit breakers (CBs) by 2020, due to the increase of the short circuit level [1]. Techno-commercial studies prove the feasibility of fault current limiters (FCL) and many investigations have been initiated and reported by EPRI and CIGRE [2] to design a practical FCL. Hybrid FCLs are among the most powerful current limiters developed so far and the most affordable idea in terms of cost [3]. In one of the hybrid FCL designs, a simple multi-contact fast switch (FS), along with the inherent features of the series arcs, has been used to commutate the current to current-limiting parallel branches [3]. Mechanical FSs can also be used in HVDC interrupters [4–6] where a rapid operation is required and semi-conductor switches are not preferred due to their high cost-loss, harmonic effects, sophisticated control, and continuous maintenance. Electromagnetic driven switches make 100-μs close/open times possible, which, in comparison with semiconductor power devices, are low-loss in ON-state and more reliable.

The actuating mechanism of many fast switch designs is based on electromagnetic repulsion. The theory of such a mechanism is thoroughly presented in [7] and the first patent filed in the late 60 s. Some more recent works on the design of such operating mechanisms have been reported elsewhere [8–10]. In a recent study, a design has been introduced to cut off 2 kA at 12 kV single-phase voltage [3,11,12], where only the mechanical model and the mechanism of FS were considered.

The electrical simulations for modeling the drive and study of the effects of different parameters on the contact speed, u_c, shows these FSs can reach $u_c = 70$ m/s through the optimum design [13], and then the arc characteristics will be severely affected by this fast elongation.

Because of the difficulties in the simulation of a fast elongating arc (FEA) in subsonic regimes, there is only a limited number of relevant papers in the literature. Some other researches were conducted on electrical simulations for modeling the drive, studying the effect of different parameters on u_c [14–16], and application of this drive for AC circuit breakers [17], in combination with vacuum CBs [18,19] or HVDC interrupters [4,10,20]. The latter has resulted in a patent registered by ABB [21]. None of these studies investigated the characteristics of FEA in FS.

Just in [22], the first 2 ms of FEA has been modeled using a black-box approach for a very limited range of arc currents and elongation speeds, without paying attention to the arc physics. Therefore, this model is only applicable to the specific geometry within a limited range of speeds and currents. It is, also, vulnerable to the problems arising in attempts to represent electric arc dynamic processes accurately utilizing ordinary mathematical models [23]. Another reason for the study of FEA is that of replacing the greenhouse SF_6 gas, where almost all environmentally friendly alternatives shall be used at pressures higher than what SF_6 was utilized at. At higher pressures, the thermal conductivity is reduced at the arc quenching temperature range, i.e., below 4–5k K, which is vital for the current interruption, and so the input energy to the arc could be reduced through the fast opening method.

The model consists of the magnetohydrodynamics (MHD), the moving mesh, and the net emission coefficient (NEC), which is the difference between the radiated and absorbed power, considering the contact effect with a circumstantial treatment of heat transfer in solid parts, including the contacts. The u_c as the output of another electromechanical finite element model (FEM) is simulated and imported into this model. Details of the electromechanical FEM model of the Thomson coil circuit, as well as technical methods for electro-mechanical measurement of fast elongated arc parameters, including the measurement noise, shift noise, and the practical way to consider their effects in simulations and experimental setup including FS geometry, are available in [24]. The first motivation of this study is to understand the physical mechanisms of arcs elongated inside the fast switches before the current zero (CZ).

A mathematical model is presented for variable-length AC arcs in contactors with elongating speeds of about 1 m/s, where the arc voltage is described by a series arc concept [25]. It is only applicable to cylindrical arcs when the temperature distribution is radial homogenous. Another already presented mathematical model for variable-length arcs tries to relate the arc voltage with the arc volume, but it is only applicable to the welding arcs when the length is changing vertically with rather low speeds [26]. All of these purely mathematical models are either modified versions of the Cassie-Mayr model and rely on energy balance inside the arc, or a modified Ayrton model for DC static arcs [27], or even a static model of welding arcs without diagonal cooling [28]. The relation between the arc length and its voltage and current has been essential, however, in other fields like arc welding, where a robust method is proposed to measure the arc length in [29]. Also, a technique to find the arc current and voltage from arc length in welding has been patented [30]. This fact can express the importance of presenting a reliable model for fast elongating arcs to solve the most critical issue related to the current commutation in hybrid FCLs/HVDC circuit breakers.

Although there are plenty of studies on the current interruption in high-voltage (HV) gas circuit breakers [31–33], vacuum interrupters [34], and medium voltage load breakers [35–37], only a few reviews on the low-speed load break switches, mostly relying on experimental findings without MHD simulations [38,39], are available. Just a few studies deal with simulations for VCB [40] and air breakers at a contact speed of about 2.5 m/s [33], which will result in a short gap before the CZ. This is the second reason for this study. In [41], the gap memory effect concerning the interrupted current is illustrated, which is more pronounced at longer contact gaps, similar to what we have here in FS.

2. Model Description

2.1. Model Assumptions

Though lots of the published investigations on MHD simulations for arcs in low-voltage miniature circuit breakers (MCB) and medium or HV SF_6-CBs are mentioned, there is almost no investigation on MHD simulation of FEAs because of some issues and simulation difficulties [42].

Increasing the fluid velocity, U, increases the ratio of advection to thermal diffusion (Peclet number). For the heat equation, this condition necessitates the use of numerical stabilization and a finer mesh. Reaching convergence on simultaneous solving of heat transfer and laminar flow (LF) equations, especially at high velocities and for gases with low viscosity, is a big challenge, as these two physical concepts are acting against each other. Sometimes, time step reduction can help, but it becomes more complicated if the generated heat increases by increasing the arc current.

Tiny meshes in the presence of a moving part result in distorted mesh elements in a few intervals. To overcome the mentioned issues and trade-off the accuracy, time, and complexity, 2-D arc simulations are used. When there are no strong vortices inside the flow, 2-D simulations are accurate enough [43]. This assumption has some effects on the gas flow that will be discussed later. In local thermodynamic equilibrium (LTE), which implies local chemical as well as thermal equilibrium [44], the plasma can be considered as a conductive fluid mixture and, thus, be modeled using the single-flow MHD equations instead of two separate flows of electron and ions. The very thin-state plasma flow can be modeled as laminar and the effects of symmetry assumption on asymmetric turbulent flows (TF) are ignored due to smaller length scales in Reynolds number (Re).

2.2. Experiment Setup and the Numerical Geometry

Figure 1a shows the elements of the prototype FS. The electromagnetic actuator of CB used in this study consists of a spiral coil in multi-layer formation, which is connected to an L-C current source through a fast-closing switch. The energy storage device consists of a capacitor bank and is connected to the current discharging coil.

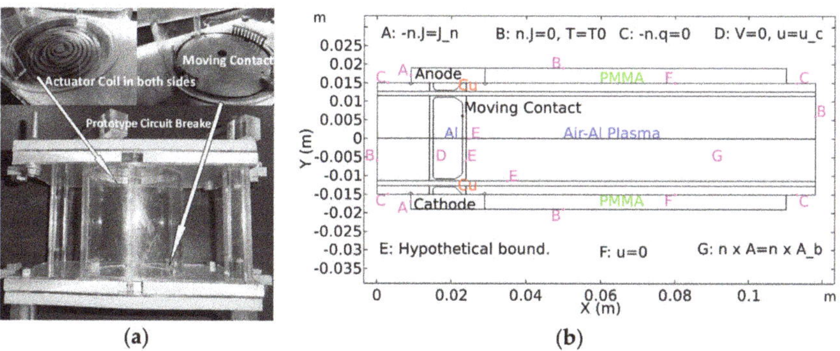

Figure 1. (**a**) Elements of prototype fast switch (FS) and Thomson coil, (**b**) 2-D geometry, and initial conditions.

Moving contact (MC) is made from aluminum (Al), while the fixed contacts are made of copper (Cu). The initial gap between the electrodes is 2 mm. The wall of the chamber is made of transparent Poly-methyl-methacrylate (PMMA), known as plexiglass. The length of the arc chamber is 11 cm, and there are two pairs of open hatches at the bottom and the top of the chamber, as shown in Figure 1a. The moving contact is accelerated by the repulsive force generated by a Thomson drive. An appropriate two-dimensional geometry of FS, including the initial conditions, is shown in Figure 1b. The gravity acts in the negative direction of the X-axis. The sharpened edges of the contacts are bent. Avoiding sharp edges is essential in solving the flow and the heat transfer equations because the sharp points generate a very high current density in terms of numerical resolution, which results in very high-temperature spots, leading to an unrealistic heat and fluid flow. The gaps between fixed and moving contact and areas around contacts are the points with intense changes in the mesh shape, as well as in the value of the heat and fluid variables. Therefore, these areas are modified by two parallel lines, and rectangular boundaries are defined around the fixed and moving contacts as hypothetical boundaries, as shown in Figure 1b. These hypothetical boundaries allow for a smaller mesh in these areas, and secondly, help to solve the moving mesh by path definition. To calculate the arc parameters in the FS model, predefined variables must be calculated in the specific regions of each arc. Therefore, another hypothetical axial boundary is defined in the center of the model (y = 0), as shown in Figure 1b, which outlines the scope of the definition of variables for two series arcs in FS. Studies have shown that 2-D axisymmetric cannot be applied, even with significant simplifications because of two reasons:

- First, the two arcs formed in FS are not symmetric. In addition to the random phenomena and the incommensurability of the two arcs in series inside the CB, the location of the anode and the cathode is opposite in two arcs, so two series arcs are entirely asymmetrical.
- Even if two arcs were symmetric, the object is not symmetric. So, the obtained solution of the axisymmetric model cannot help to determine the shape and length of arcs. Therefore, the model must be defined in Cartesian mode (2-D).

2.3. *Properties of the Material*

2.3.1. Gas (Air-Al Plasma)

Thermal plasma properties of air-Aluminium vapor mixtures in [45] show that even a small amount of metal vapor at atmospheric pressure has an appreciable influence on the radiation and the electric conductivity (σ), but a negligible effect on the other features. The effect of metal vapors on σ is dominant in low-current arcs while the impact on radiation dominates at high currents [46] due to higher temperatures (T).

Plasma conductivity in the presence of 1–3% metal vapor in 11,000–13,000 K, is about 20–30% higher than of a pure air plasma [47,48], which is modeled here. When the current is lower than 1 kA, metal vapor resides only in two small regions (1–3 mm) in front of the two electrodes because of limited electrode vaporization [46,49]. Consequently, the arc temperature drops significantly in these small regions near the cathode [50,51]. Then, arc voltage (V_{arc}) in low current mode will not be sensitive to the presence of metal vapor [49,51].

If absorption is ignored (thin layer) then metal vapor increases the NEC, but if absorption is considered then the presence of Al vapor despite Cu reduces the NEC in the temperature range of 10–15k K, if the Al mixture ratio is less than 5% [45] and arc radius is less than 5 mm, which applies to our study. The NEC decreases by an increase in the arc radius [52]. The reported results are utilized to obtain the thermodynamic properties of hot air, including heat capacity, viscosity, density, thermal and electrical conductivity, as well as NEC for the temperature range up to 25,000 K at constant atmospheric pressure [53,54] and is calculated for higher pressure (20 bar). Although the exact formulation of the energy transport by radiation is very complicated, the three mostly used methods are P1, method of partial characteristics, and NEC.

A comparison between these methods is discussed in [55]. The NEC has acceptable accuracy and less entanglement in the range of the optical thicknesses and arc temperatures, (10–15k K and 2–6 mm) [56]. The Air-Al mixture and NEC absorption effect in the presented model will be explained here.

The core diameter varies along with its length at different times. This variation is shown in Figure 2a.

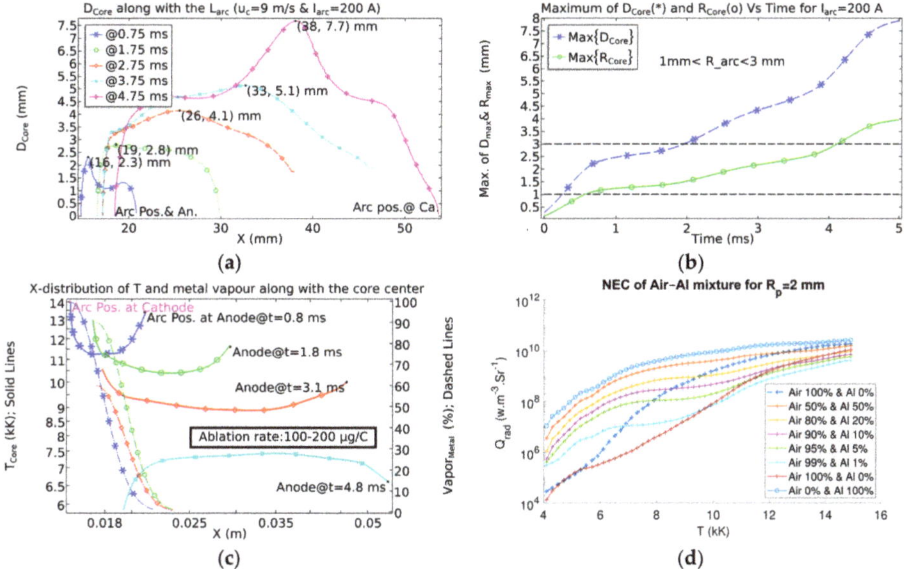

Figure 2. (**a**) Variable core diameter along with its length, (**b**) Maximum of core radius, (**c**) Distribution of T and metal vapor along the core center, and (**d**) net emission coefficient $NEC_{air\text{-}Al}$ in arc chamber for 200 A_{peak}.

As it is shown in Figure 2b, the maximum radius of the core varies between 1 mm and 3 mm. So, the radiation absorption radius, R_p, for the NEC model is assumed 2 mm. The ablated mass for Cu electrode in low currents is calculated around 100–300 µg/C [57] and in our stationary study was around 316 µg/C [51]. Although the melting point of Al is half of Cu, its latent heat of evaporation is twice the Cu. Assuming a similar ablation rate will result in 150–300 × 10^{-3} cm^3 Al vapor. By utilizing the particle tracing method [51], the ratio of Air-Al is calculated and it is shown along with the core center in Figure 2c, which complies with other studies [49]. According to [58], the influence of changing the arc radius from 1 mm to 10 mm on NEC for air plasma in the temperature range of 7–15k K is not significant, and its gradient by increasing the arc radius is negative. So, finding an average radius for the arc gives an accurate estimation of NEC. But according to [45], the influence of metal vapor percentage is enormous at this temperature range. So, metal vapor has been considered, and the arc radius was averaged. Based on the reported NEC of air-aluminium mixture [45] and temperature distribution in Figure 2c, NEC in the arc chamber is calculated through recursive studies to minimize the relative error.

The result is shown in Figure 2d. The volumetric proportion of Al-vapour in a 2-mm gap between electrodes is more than 90% at the start of the arc. Particles have Maxwellian speed distribution, and contact is moving, which results in FEA. So, particles disperse along the arc length, and the mixture ratio decreases as time elapses. Most of the Cu particles remain near the fixed cathode while Al particles are dispersed by moving cathode displacement, as velocity inside the core is around 50 m/s at 0.5 ms but falls to 5 m/s at 1.5 ms. So, the mixture ratio of Al falls to below 5% in an arc core.

The main parameters of the utilized model are the transport and thermodynamic properties of gas mixtures, which would be obtained from the articles or could be calculated through chemical physics [59,60] for pressures and mixtures other than what is reported in the articles. Software packages handle these chemical physics calculations. Here, we used the PLASIMO® package [61], and the electrical conductivity, total heat conductivity, mass density, NEC, viscosity, and heat capacity of pure air and air-Al mixture at 1 and 20 bar at different mixture ratios and pressures are presented in Figure 3. The legends for all sub-figures are the same as Figure 3a.

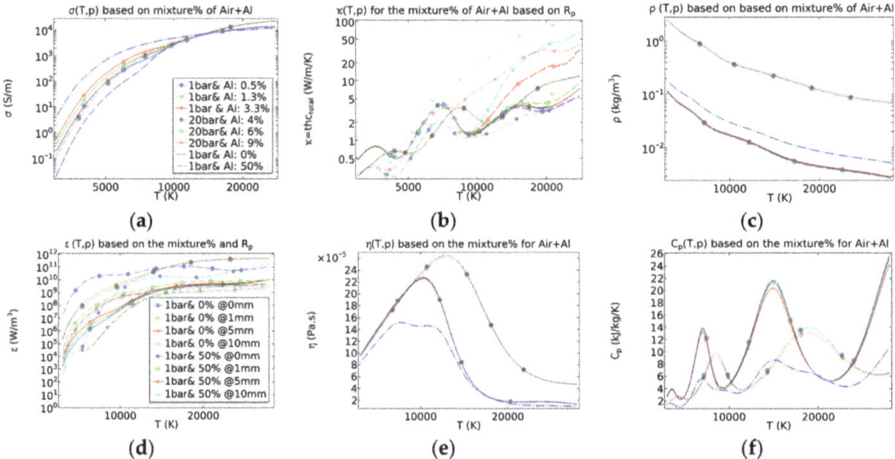

Figure 3. (**a**) Electrical conductivity, (**b**) Heat conductivity, (**c**) Mass density, (**d**) NEC, (**e**) Viscosity, and (**f**) Heat capacity of pure air and air-Al mixture at 1 and 20 bar at different mixture ratios.

The changes in the pressure are very small in this study so its effect on vaporizing temperature is ignorable, but it could be considered as it was explained in the above through assuming a two-phase thermodynamic system and utilizing the method explained in [62].

2.3.2. Solids

The necessary parameters for the solid parts, including Al [63] and Cu contacts, as well as PMMA [64–66] enclosure to be considered in the numerical model, are presented in Table 1.

Table 1. Required parameters for modeling of solid parts.

Property	PMMA	Al	Cu	Unit
C_P	1420	900	385	J/(kg·K)
κ	0.19	238	400	W/(m·K)
σ	1×10^{12}	3.774×10^7	5.998×10^7	S/m
ρ	1190	2700	8700	kg/m^3

2.4. Equations, Initial, and Boundary Conditions

2.4.1. MHD Equations and Sheath Model

In general, the MHD method is time-consuming but with less data acquisition costs, while many rapid mathematical models do not care about the data acquisition costs. So, some rapid models are full of coefficients, the application of which is inevitably condemned to death after an ad hoc implementation. Therefore, it is essential to define a model description as much as possible on the data flow and information, which is the case of the MHD simulation where a mathematical model that stands on such a solid basis will have a high probability of being applied, and so tested and improved, in a continuous way [67]. That is the reason for the vast utilization of the MHD tools for the design and performance evaluation of switchgear [68]. Four groups of equations used in this study are the current conservation equations Ampere's law and Lorentz force, HT, and Navier-Stokes equations (NSEs) for an incompressible fluid, including conservation of momentum and mass. Details of equations were already presented for the fixed contact simulation [51]. In Newtonian fluids (all gases and most non-viscous liquids [69]), if the Mach number (*Ma*) is more than 0.3 or the *T* inside the fluid changes significantly and rapidly, the fluid is considered compressible. In our study, although the *T* rises sharply, the changes are not in a vast range, and calculation shows that *Ma* is not more than 0.3. Also, the Re is relatively small, so the flow is laminar in each layer [51]. The *T* in the core is in the range of 9–15k K and densities of ions and electrons at this temperature are high enough to consider that NSEs are valid for highly ionized air plasma [51]. This assumption is justified by the high plasma pressure and the large estimated electron densities in the order of 10^{18} cm^{-3} [44]. Here, the sinusoidal 50 Hz 200 A current source $J_n(t)$ is applied to the terminals, as shown in Figure 1b.

$$J_n(t) = \pm 5 \times 10^5 \cos(0.314 \times t[\mathrm{ms}^{-1}])[\mathrm{A/m^2}] \tag{1}$$

The initial conditions for these equations are zero potential at moving contact and zero current density at the outer wall of ACH, as shown in Figure 1b. Due to the current injection direction in the quarter-cycle simulation, the lower side of the moving contact and the upper fixed contact act as an anode, while the lower fixed contact and the upper side of the moving contact are the cathodes. The required parameters to model the effect of the Al and Cu cathode and anode are obtained from [66,70]. To solve the heat transfer equations together with the magnetic field equations in a 2-D domain, the thickness should be considered, which is added as d_z = 10 mm compared with 2-D axisymmetric equations based on arc diameter.

$$\rho C_p \frac{\partial T}{\partial t} + \rho C_p u \cdot \nabla T - \nabla \cdot (k \nabla T) = Q_{tot} \tag{2}$$

As it is shown in Figure 1b, the top and bottom of the enclosure, interior arc chamber, as well as the outer surface of the walls are modeled with the ambient temperature T_0, which meets the actual status. It is also assumed that heat transfer from the open points to the outside is negligible. Therefore, these hatches are modeled as thermally insulated surfaces ($\vec{n} \cdot q = 0$).

As the time steps and solution methods to solve the MHD and electrodynamic drive are different [71], an appropriate approach is an asynchronous solution. Measurements, and electrodynamic drive simulations, show that moving contact reaches a constant speed at the start of the arc [24]. So, the simulated speed from the electromechanical solution is imported into the fluid flow model as $\vec{u}_c$. The influence of thermionic electron emission (3–7) is not significant on the current density (*J*), as at the highest melting point of Cu cathode 4.16 × 10^{-5} A/m^2 at *T* = 1356 K, while *J* at the cathode is in the order of 9.3 × 10^8 A/m^2. Increasing *T* does not ensure observable electron emission since Cu will melt and then evaporates or decomposes, but it affects the contact temperature by keeping the continuity in heat transfer equations. Although the first studies on liquid metals [72] show an explosive electron emission from liquid cathodes in special conditions and for a limited time, as far as we know thermionic emission data exist only for polycrystalline solid-state copper emitters [73] and Cu is not the ideal thermionic emitter due to its relatively low melting point (1356 K), which limits the working temperature region to 800–1100 K or so [74]. The cathode temperature does not reach the melting point in the first cycle, so thermionic emission from melted Cu is ignored.

$$\vec{J}_R = A_R T^2 \exp\left(-\frac{q\Phi eff}{k_B T}\right) \cdot \vec{n} \tag{3}$$

$$J_{ion} + J_{elec} = \left|\vec{J} \cdot \vec{n}\right| \tag{4}$$

$$J_{elec} = \begin{cases} \left|\vec{J} \cdot \vec{n}\right|, & \left|\vec{J} \cdot \vec{n}\right| \le J_R \\ J_R, & \left|\vec{J} \cdot \vec{n}\right| > J_R \end{cases} \tag{5}$$

The required parameters of Cu contacts are ϕ_{eff} = 4.94 V, and A_R = 120.2 A/(K·cm)^2 [66,70]. According to (6), the cathode temperature change is the result of the energy absorption and electron emission from the cathode, i.e., overcoming the space charge energy level, and the potential of the plasma ionization layer of positive ions (V_{ion} = 15.5 V) near cathode regions [75].

$$-\vec{n}\cdot(-k\nabla T) = -\vec{J}_{elec}\Phi_{eff} + \vec{J}_{ion}\cdot V_{ion} + J_e\cdot\Phi_{cathode}\left(\vec{J}^{max}_{nearC}\right) \tag{6}$$

According to (7) at the anode surface, the temperature change in the anode is a result of the absorption of electrons at the anode, i.e., overwhelming the energy level because of the accumulation of electrons near the anode.

$$-\vec{n}\cdot(-k\nabla T) = \left|\vec{J}\cdot\vec{n}\right|\cdot\Phi_{eff}\left(\vec{J}\right) + \vec{J}_{e}\cdot\Phi_{anode}\left(\vec{J}^{max}_{nearA}\right) \tag{7}$$

$\Phi_{cathode}\left(\vec{J}^{max}_{nearA}\right)$ and $\Phi_{anode}\left(\vec{J}^{max}_{nearC}\right)$ are voltage drops (V_d) as a function of J over the fluid–contact interface without taking any physical aspect of the sheath into account [76].

2.4.2. Moving Mesh and Mesh Validating

As already noted, another simulation difficulty is fast movement. To consider the mechanical motion, Plexiglas walls and fixed contacts are modeled with fixed mesh, but the rest of the interior of FS is defined as a region with free deformation, and a mathematical lema named moving mesh [77] is used to model contact motion. The mesh around the contact moves with u_c. The maximum and minimum sizes of mesh elements were selected as 1.06 mm and 15.2 µm, respectively, but as it is shown in Figure 4b, boundary layers of fixed contacts, as well as the edges of the fixed and moving contacts, are modeled by a fine mesh with a minimum/maximum size of 0.76/70 µm suitable for CFD.

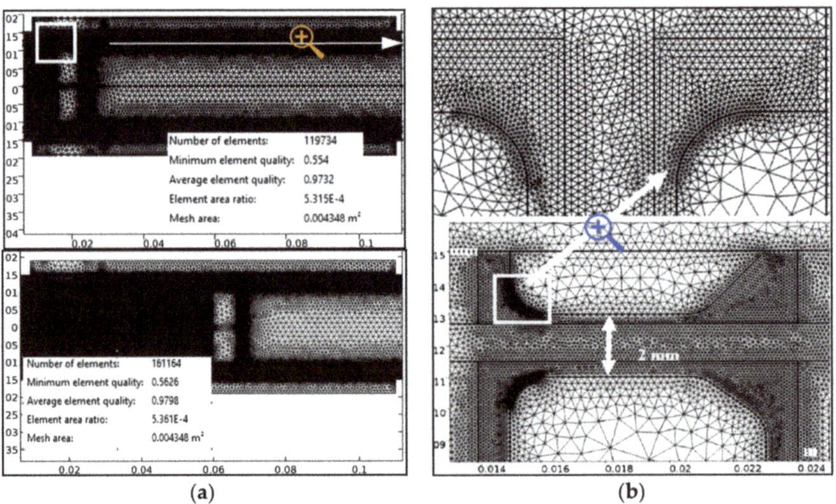

Figure 4. (**a**) Moving mesh with 119,734 elements in 43.48 cm² at first and 161,164 elements at the end and (**b**) Extremely fine mesh defined around contacts and gap.

The areas between the sidewalls and the bottom of FS, as well as the linear boundaries parallel to the walls, are modeled by a fine triangular mesh with a maximum size of 150 µm.

To perform a more realistic fluid flow simulation next to the walls, a boundary layer mesh on all internal walls is defined. The final mesh has around 120,000 elements, as shown in Figure 4a. The mesh element quality is shown in Figure 4a is a dimensionless quantity between zero and one, which is an important aspect when validating a model. This measure is based on the equiangular skew.

It penalizes elements with large or small angles as compared to the angles in an ideal element. Poor quality elements are considered for quality below 0.1 [78]. Contact movement causes the meshes to be squeezed and drawn, resulting in element quality reduction. Massive compression and stretch cause the model to be unstable and unsolvable. If the mesh overlapping reaches a predefined level, the solution is stopped, and the mesh is repeated considering the previous solution as the initial condition for the new meshed geometry.

In this way, an automatic mesh is generated almost every 120 μs up to 44 times in the course of the simulation, so that the output of mesh number 45 according to Figure 4a contains about 161,000 elements in 43 cm^2 with the same quality as the first mesh. High mesh density below and around moving contact, as well as next to the walls, is visible. Taking into account the equations modeled with this number of mesh elements leads to 460,000 degrees of freedom (DoF) in equations for the primary mesh, which finally reaches more than 635,000 DoF by increasing the number of meshes. The minimum mesh quality that has been ensured for each of the models that have been solved is 0.55, with an element area ratio of about 5.3×10^{-4}. In general, the triangular mesh is a quick and straightforward way to obtain meshes of high element quality but comes with a diffusion cost. However, the extra diffusion can sometimes be desired, since it becomes easier to achieve convergence. A precisely defined triangular mesh specifying the distribution of mesh elements along an edge, in addition to Refinement of the Corner to decrease the element size at sharp corners, is perfect for our model, where the mesh serves as a starting mesh for mesh adaption. The obtained solution is then used to refine the mesh based on some indicator function. Finally, the adapted mesh is used to simulate the time interval again. After examining different methods, the parallel sparse direct and multi-recursive iterative linear solver (PARDISO) [79], as a stable time resolution method, was used while the Jacobian matrix was updated at each step. Absolute tolerance of 5×10^{-4} and at least 1 ps time steps at the start of the solution and maximum 2 μs is vital for modeling of V_{arc} in the last 20 μs of the arc cycle accurately.

2.5. *Arc Ignition*

To initiate the arc in the most of the other researches, a predefined conducting column, usually rectangular with a temperature between 7000 and 10,000 K, is assumed between two contacts, which has the role of a thin melting wire in practical experiments [71,76,80], but it causes the shape of the arc and its movement to be unrealistic during the initial arcing period. As an accurate arc shape and therefore V_{arc} in the first 200 μs is necessary for current commutation purposes, a different approach has been taken here. The initial air conductivity is assumed to be 1 s/m wherever its conductivity is less than 1 s/m. By taking this approach along with electron emission from contacts, the arc starts from the sharp points and with an accurate initial shape. Then, the arc region can be determined using the σ distribution.

2.6. *Turbulent vs. Laminar*

2.6.1. Plasma Numbers

Critical Re must be determined through experiments or numerical simulations for each configuration.

The maximum of *Ma*, Peclet number (Pe), Re, and Prandtl number (Pr), calculated in each cell of meshes for the whole of simulation time in an arc of 200 A, are shown in Figure 5. *Ma* is much less than 0.3. Pe, the product of the Re and Pr, is very large. The studied case is, however, similar to the flow inside a cylinder, where for $41 \leq \mathrm{Re} \leq 10^3$ an unsteady but predictable LF with counter-rotating vortices shed periodically from the cylinder is observed [81]. If $\mathrm{Re} > 10^3$, vortices will be unstable, resulting in a turbulent wake behind the cylinder that is 'unpredictable' [81]. There is a transition to an entirely TF. As the stream begins to transfer to the turbulence regime, fluctuations appear in the flow, even though the inlet flow rate does not change with time. Then, it is no longer possible to assume time-invariant flow. So, it is compulsory to solve the time-dependent Navier-Stokes (TDNS) equations.

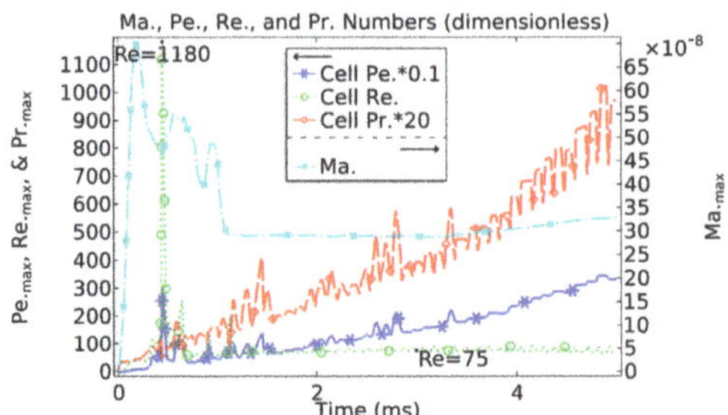

Figure 5. Maximum of cell Peclet number (Pe), Mach number (*Ma*), cell Reynolds number (Re), and cell Prandtl number (Pr) for the whole of simulation in 200 A_{peak} arc.

2.6.2. Turbulent Model

From Figure 5, Re becomes more significant than 10^3 at 0.5ms. This is an indication of TF. So, the algebraic *Y-plus* TF model is used in the first 500 µs of the simulations. Figure 6 shows the arc temperature distributions at different times. The application of the TF model results in lower temperatures and a different shape for arc between contacts (Figure 6d).

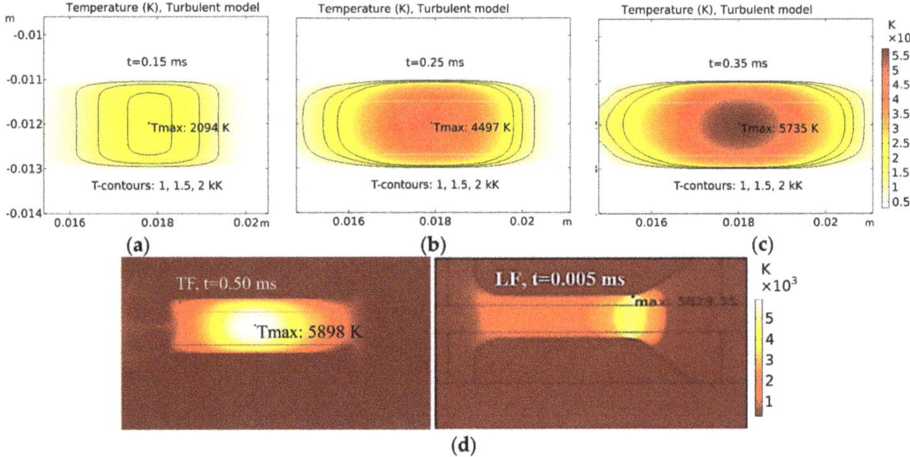

Figure 6. Arc temperature surface modeled in turbulent flows (TF) at (**a**) 0.15 ms, (**b**) 0.25 ms, (**c**) 0.35 ms, and (**d**) Comparison of arc shape in the laminar and TF model for an arc of 200 A_{peak}.

No large vortices field occurs inside the arc due to a border between the thermal plasma column and TF around it [82]. So, if the mesh is fine enough to resolve the size of the smallest eddies in the flow, the laminar model will remain suitable to reduce the time costs.

2.7. The Arc Roots and the Sheath Model

The simulated temperature distribution in the arc core and the contacts at two different times, 250 µs and 1 ms, are shown in Figure 7a to identify the dominant physical processes inside the arc during different phases of arcing. Up to 250 µs, the arc is sliding in the opposite direction of the moving contact (MC) displacement due to the cold air blowing at the arc column, (see Figure 7c). As it is shown in Figure 7a, in the first 250 µs, the arc length is constant, and the temperature distribution is almost homogenous inside the core, except for the anode and cathode regions. Therefore, the arc-elongating speed is zero before 250 µs.

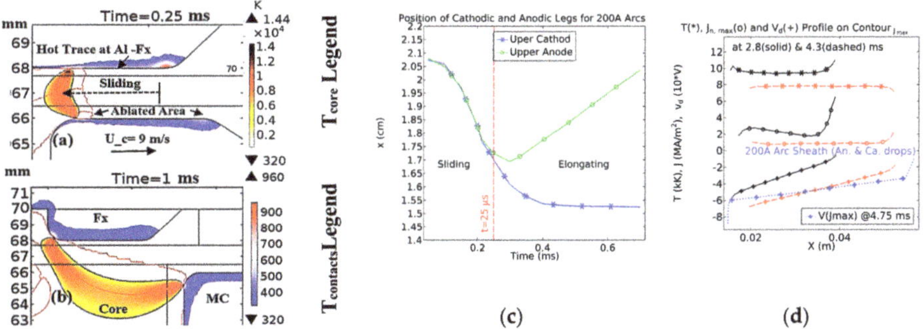

Figure 7. A Core and contacts' temperatures show (**a**) Arc sliding along the contacts up to 0.25 ms, (**b**) Arc elongating between fixed (Fx) and Moving (MC) contacts at the speed of 9 m/s, (**c**) X-position of arc roots vs. time, and (**d**) Temperature, current density, and voltage profile along the contour of J_{max}.

The arc is not in LTE in these two regions, and the NSE condition is not fulfilled. But simulated results are not so far from the measurements [24]. As it is shown in Figure 7a, arc sliding causes a hot trace at the contacts, which have a temperature range of 600–900 K based on $T_{contact}$ legend. As it is shown in Figure 7a, the arc temperature is higher than 10k K, i.e., the air is highly ionized, and the NSE assumption is valid. From 1 ms, as it is shown in Figure 7b, the arc is elongating between two points at the contacts. The arc root model is based on [76,83] and the temperature, current density, and voltage profile along the contour of J_{max} in Figure 7d show the effect of this model.

3. Results

3.1. Temperature and Conductivity Dynamics

Figure 8a,b shows σ contours and the point of the maximum conductivity at 1.35, 2, 2.75, 3.45, and 5 ms in an arc of 200 A. The bright-dark part (green contour) is named core of arc. The rest of the arc inside the red contour is named arc column. The arc light is due to radiation that is proportional to T^4. So, the arc column boundary is defined as $\sigma = 1$, which is equal to 3300 K but the core is the gliding section of arc ($\sigma > 600$, which is similar to 6700 K in air temperature) which is passing more than 80% of the arc current.

Figure 8 shows contours of (a) σ, and (b) the temperature gradient at 1.35, 2, 2.75, 3.45, and 5 ms in an arc of 200 A. It shows that before 2.8 ms the temperature gradient inside the core is high, but after this time it becomes more uniform inside the core center. The point of maximum temperature (T_{max}) and maximum conductivity is clear in the figures. It is a point with the same color as the related text at the upper left of it. It is close to the moving anode and then shifts to moving cathode but before the CZ falls to the core midpoint. Figure 8d shows that after 2 ms of arc ignition and independent of arc current magnitude, the core is passing 70–80%, and the column is passing around 20–30% of current. The axial centerline of J in core shown by the red line is where the J gradient perpendicular to the arc column is zero. It is not the spatial centerline, so the core shape and thickness on both sides of the arc center line are not similar due to massive gas flow to the outer sides. Figure 8e shows zero gradient contours for the normal component of current density J_{norm} (green), σ (red), and temperature (blue) at 1.4 ms. Due to the relatively linear relationship between temperature, σ, and current density in the range of 10–14k K, these contours overlap.

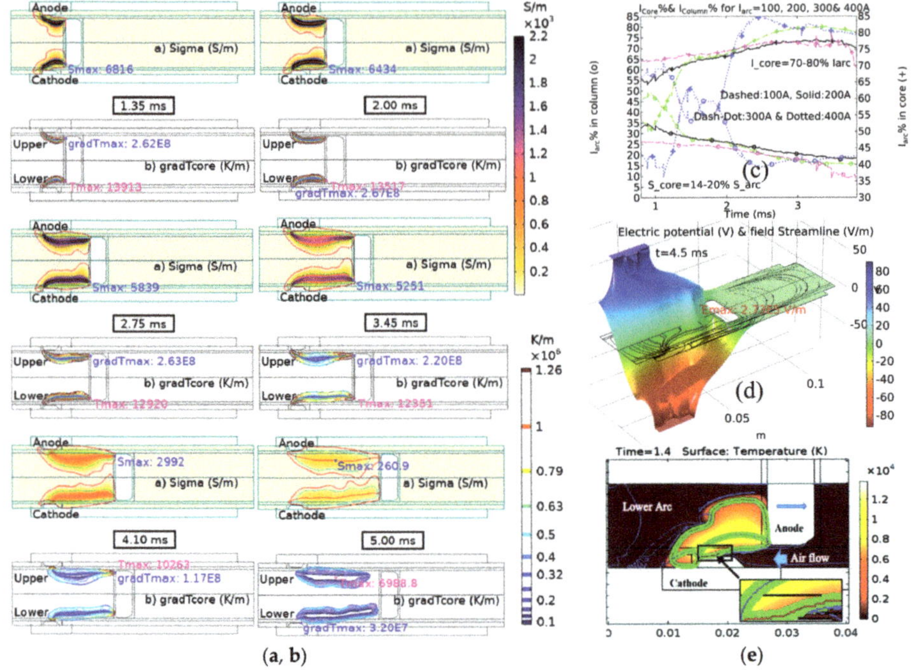

Figure 8. (**a**) σ contours and the point of maximum σ, (**b**) Contours of T-gradient at 1.35, 2, 2.75, 3.45, and 5 ms in an arc of 200 A_{peak}, (**c**) Percentage of $I_{arc} = I_{source}$ divided between core/column, (**d**) Electric potential and electric field at 4.5 ms showing anodic and cathodic V_d, and (**e**) Zero gradient contours for J_{norm} (green), σ (red) and T (blue) at 1.4 ms on the T-surface.

3.2. Simulated vs. Captured Appearance

Figure 9a displays time-tagged pictures from simulated results of the arc border (outer contour) and core border (inner contour) based on the above definition, including the point of T_{max} in 200 A_{peak} arc (upper left of T_{max}). Figure 9b illustrates the arc imaging in FS taken by a 15,000 fps recorder. A dark green filter is used at the camera to prevent saturation, so pictures before 1.57 ms and after 3.37 ms are not so visible. The visual appearance of the curved and the other pulled simulated cores and the core brightness at 2.8 ms (before the change in mode) are confirmed by the experimental results in Figure 10b. Arc current is calculated from:

$$i(t) = I_{peak} \times \cos(0.314 \times t[\frac{1}{\text{ms}}]) \tag{8}$$

A low-frequency (f $\leq$ 100 Hz) arc on each half-cycle mirrors a direct current arc, the contact separation initiates the arc, and when the current reverses the electrodes interchange their roles as anode or cathode [84]. Simulated and measured arcs are not at the same cycles, so electrode positions are reversed. From Figure 9a, the arc starts with a T_{max} of 14,790 K on both sides of FS from the sharp points of the contacts. Then it moves downward along the contacts and stays at its bottom because of the intense flow on it. When the arc comes out of the gap between the contacts, its T decreases up to 1000 K after 0.25 ms. Then it is elongated, but before 3.6 ms, its temperature is not reduced so much despite the current falling according to Equation (3), and it remains constricted. It is observed that the core is highly compact, hot, and luminous before 2.95 ms, and the arc boundary is more spacious than the core border.

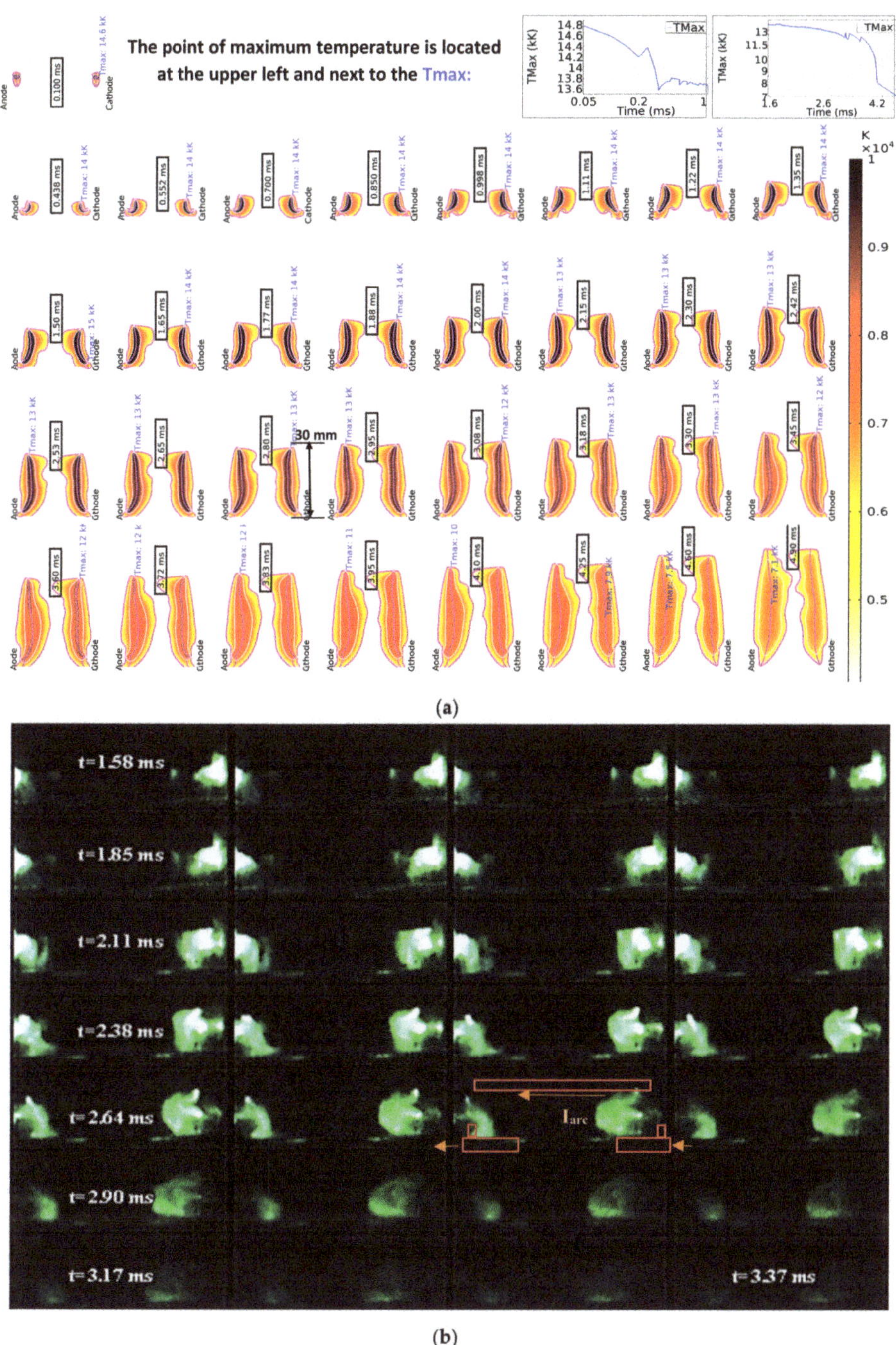

Figure 9. (**a**) Time-tagged simulated results of arc and core boundaries for two arcs in series, point of T_{max} in 200 A_{peak} arc, and (**b**) High-speed imaging with 15,000 fps.

This mode is obvious from high-speed imaging with 15,000 fps in Figure 9b and we call it a constricted mode. At 3.83 ms, the T dropped sharply. The position of T_{max} has been recorded near the moving contact before the mode change, but later it moves to the center of the arc at 4.25 ms. That is, the arc starts to cool on both ends, but the current is not zero yet. We refer to this as a dispersed mode.

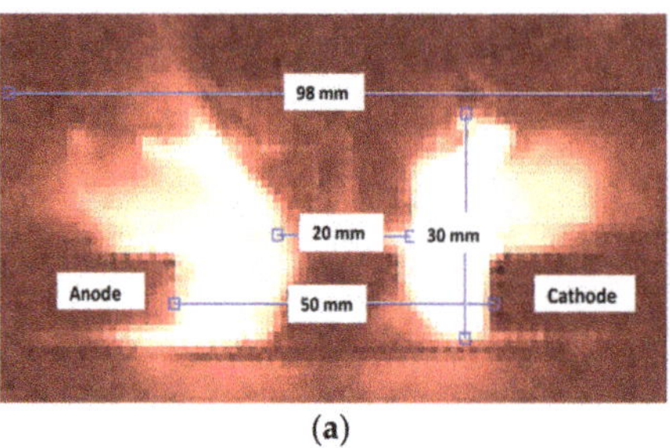

(a)

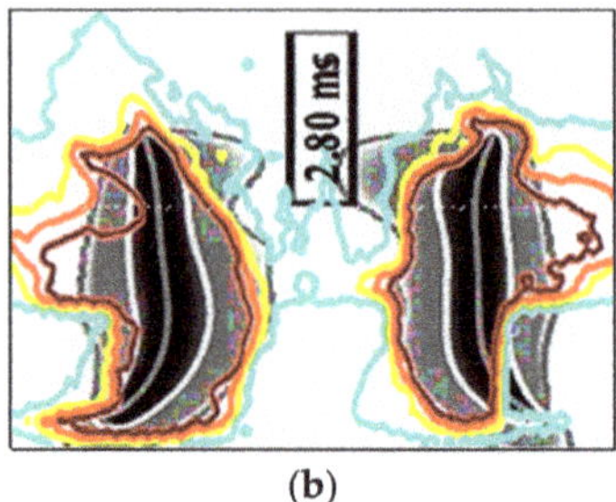

(b)

Figure 10. (**a**) Enlarged image of the arc and (**b**) Its light intensity contours extracted by the image processing vs. simulated arc at 2.8 ms.

An enlarged and horizontally mirrored image of the arc at 2.8 ms is shown in Figure 10a and the light intensity contours [85] extracted by the image processing technics [24] of Figure 10a overlaid on the simulated one are shown in Figure 10b.

3.3. Impact of Thermionic Emission

Thermionic emission acts as a heating source, and it is of crucial importance for contact temperature and ablation simulations, especially for the Cu cathode. The effect of modeling the electron emission from the cathode on the T_{max} of contacts is compared in Figure 11. "C. Eff." refers to the contact effect and means considering the thermionic emission in simulation. Considering thermionic emission, the temperature of the moving contract does not differ significantly, but as it is shown in Figure 11a the T_{max} of the fixed cathode increases at 3.4 ms from 707 K in Figure 11b to 1103 K at 1.8 ms, while the T_{max} of the fixed anode increases from 660 to 800 K. Contact erosion mostly happens near the fixed Cu cathode and then moving Al anode. It was modeled in detail and reported in other recent research [51] and it is also clear from Figure 12a.

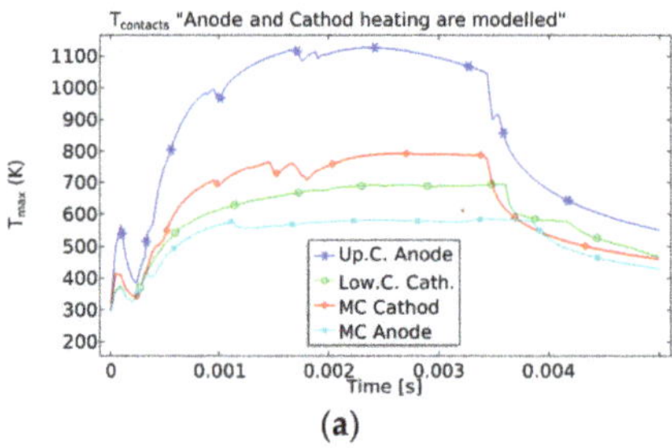

(a)

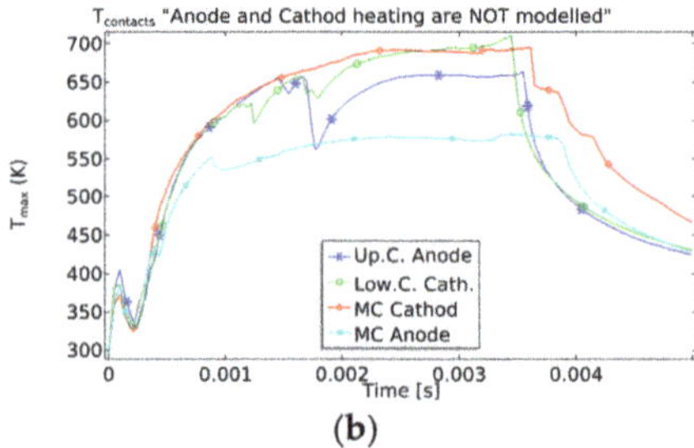

(b)

Figure 11. T_{max} of contacts (**a**) with thermionic emission modeling, (**b**) Without thermionic emission modeling in 200 A_{peak} arc.

Figure 12a compares the simulated voltage for two 200 A arcs with and without thermionic emission modeling. It reveals that the contact modeling in the first 3.5 ms has no significant effect on V_{arc} for two 200 A arcs, but it affects the voltage in the last 1 ms. Anodic and cathodic V_ds are modeled in both simulations. Figure 12b shows the melted/ablated points on fixed and moving contacts of the prototype switch. As predicted by the simulations, a trace of arc root is visible on the fixed cathode contact, and the most significant erosion happened at the lower part of the fixed contacts.

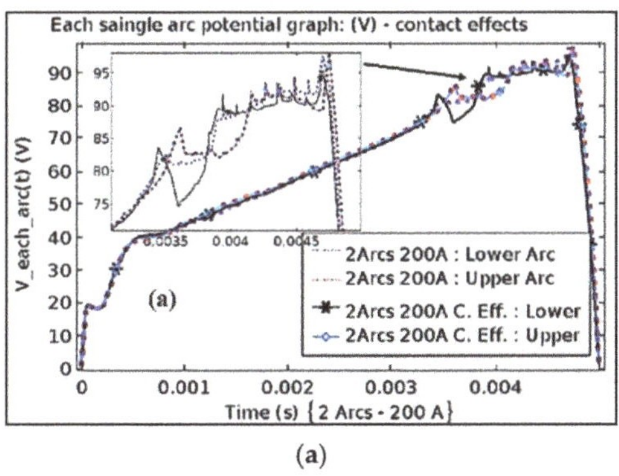

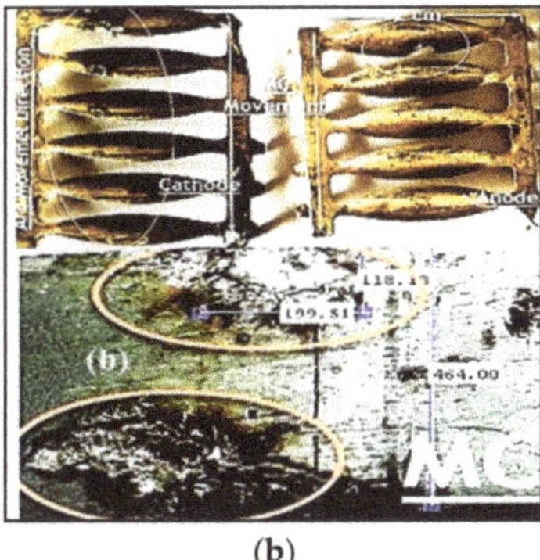

(a) (b)

Figure 12. (**a**) Ablation on fixed and moving contacts (MC in figures means moving contact) and (**b**) Two 200 A_{peak} arcs with(out) thermionic emission modeling.

3.4. Arc Mode Change and Interaction of Arcs in Series

Figure 13 shows the influence of the arc mode change and its effect on increasing the core area. It shows the instantaneous area of the whole arc and its core, as well as the ratio of core-to-column area, for 100 and 200 A arcs with(out) thermionic emission modeling. The slope of the ratio increases suddenly by a factor of 2.15 for 200 A arcs between 2.95 to 3.83 ms, which is the time of change in the arc mode and means the core is not constricted any more. A single 100 A arc is simulated by applying $J_n(t)$ to one of the fixed contacts. The arc area is smaller, but mode change happens in lower currents.

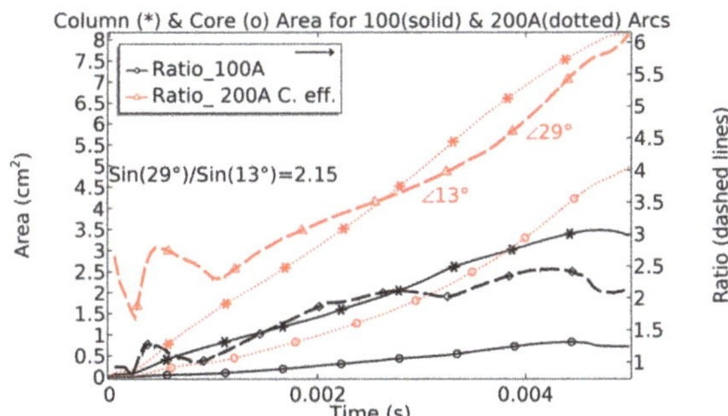

Figure 13. The core area, and the ratio of core-to-column area, for 100 and 200 A_{peak} with(out) thermionic emission.

4. Discussion

4.1. Cathode and Anode V_d

Figure 14a shows arc boundaries, the σ inside the core (yellow), Electric potential (V) contours (red and blue contours perpendicular to arc and core boundaries), arrows, and contours of J_{norm} in logarithmic scale (in green). The numbers in black are the V-contour tags in 2.5 V steps between. A zoomed view of conductivity for the 1 mm-thick layer near the cathode is shown at the upper right corner of Figure 14a and in Figure 14b for the upper and lower sides of the moving contact. The σ_{max}, T_{max}, and consequently J_{max} are evident in this view, and numbers in magenta are the J-contour tags.

The arc column gets narrow near the cathode, while it has a bezel shape near the anode. From Figure 14c,d, it is obvious that the arc core (yellow section) has a sharper tip near the cathode than the anode.

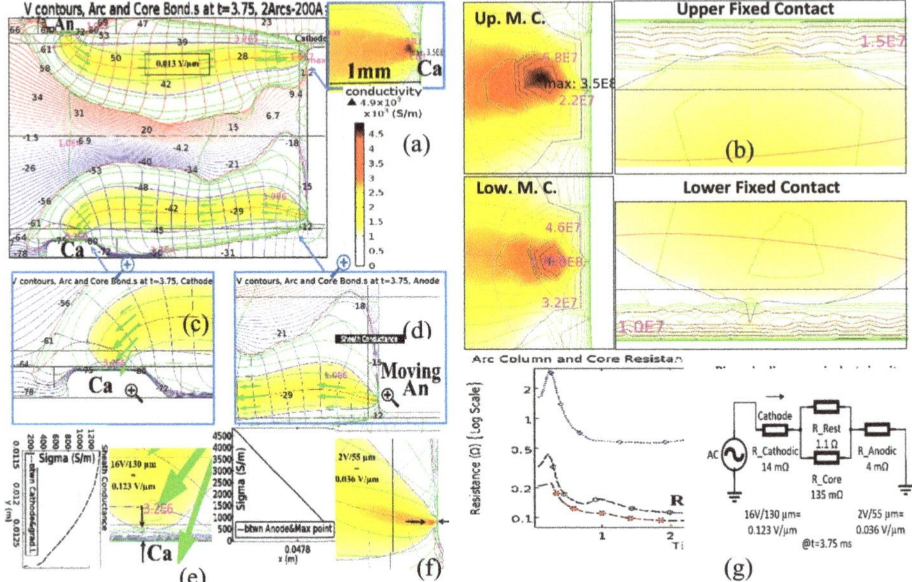

Figure 14. (**a**,**b**) Arc boundaries, V contours, and J_{norm} in Log. scale and the zoomed view of the near electrode areas at (**c**) Fixed cathode, (**d**) Moving anode, (**e**) Diagram of $\sigma_{sheath}(x,t)|_{t=3.75\ ms}$ from cathode to the core gradient line, (**f**) Diagram of the $\sigma_{sheath}(y,t)|_{t=3.75\ ms}$ from anode to the σ_{max}, and (**g**) Simulated arc and core resistance and piecewise linear equivalent circuit of $R_{arc}|_{t=3.75\ ms}$ for 200 A_{peak}; (An, Ca, and MC in figures refer to the anode, cathode, and the moving contact).

The zoomed view of these tips is shown in Figure 14e,f, showing the sheathes. The diagrams of the $\sigma_{sheath}(y,t)|_{t=3.75\ ms}$ from cathode to the core centerline and $\sigma_{sheath}(x,t)|_{t=3.75\ ms}$ from anode to the σ_{max} are shown in this figure. The equipotential lines between the core and the cathode wall are very dense, which means a rather uniform field and significant linear V_d in the cathode sheath; this finding is shown in the conductivity curve of Figure 14e and is in line with other studies [86]. The sheath effect increases the arc resistance (R_{arc}) in front of the electrodes, as it is recognized from the broken and deformed σ diagram of the cathode sheath in Figure 14e.

It shows a change in σ (blue) near the sheath layer and the constricted V contours (brown) near the electrodes. The J at the cathode depends on cathode material, while it is independent of the current [87]. The J_{max} near the moving cathode was simulated to 3.5×10^8 A/m^2. The relation between J and electron temperature is presented in [75], and the sheath thickness is estimated based on electron temperature [88].

Our simulated sheath thickness is about 130 μm at the cathode, which complies with the mentioned research. At 3.75 ms, the simulation shows about 16 V drops along the 130 μm cathode sheath (123 mV/μm) and about 2–3 V drops in 55 μm (45 mV/μm) between the core boundary and anode. Simulated anodic and cathodic voltages are similar to others [89].

The thickness of electrode sheaths for arcs at atmospheric pressure is less than 0.7 mm in total. The sheath V_d and the current divided between core/column imply a piecewise linear equivalent circuit of $R_{arc}|_{t=3.75\ ms}$ for 200 A_{peak}, shown in Figure 14g with the simulated arc and core resistances.

4.2. Effects of Fast Elongation on the Arc Voltage

Figure 15a compares the measured V_{Meas} (dash-dotted in green) with the simulated $V_{simulated}$ for 200 A, considering thermionic emission modeling. It is seen that both V_{Meas} and $V_{simulated}$ jump from zero to 24–33 V, then step up to 40–45 V and then increase to about 75–80 V with a linear increment rate of m = 10 kV/s.

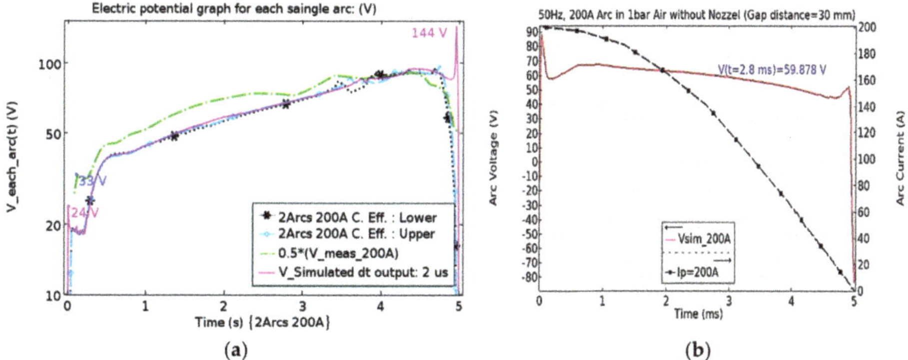

(a) (b)

Figure 15. Measured and simulated arc voltages for 200 A_{peak} arcs in (**a**) Prototype FS with u_c = 9 m/s and (**b**) Fixed contact distance of 30 mm initiated by exploding wire at the same current cycle.

The shape of V_{arc} complies with other researches [90], but the first jump of the measured voltage is higher because of higher simulated temperature in laminar modeling. The jump at the end of V_{arc} is missed in measurement and simulation with a sampling rate of 50 μs, but it is detected with a sampling rate of 2 μs (in purple) of simulated output and is apparent in other experiments [37,91,92]. The slope of the measured and simulated voltage falls close to zero with the arc mode change. Figure 15b shows the simulated V_{sim} for 200 A_{peak} arc between a fixed contact distance of 30 mm (stationary arc) initiated by exploding wire at the same current cycle [93]. V_{arc} is higher in the elongated state.

Figure 16a compares the measured (Dashed Line) and simulated (Solid Line) V_{arc} for 200 A arc. The higher the u_c, the faster the elongation and the higher the voltage for the similar arc currents. By increasing u_c to 22.5 m/s, V_{arc} reaches six times its value in the fixed contact distance at 200 A. Figure 16b shows the influence of elongating speed on V_{arc} for 200 A arc. The m [kV/s] is related to the u_c, thus it increases in higher u_c. The V_{arc} is an essential measure for a successful commutation.

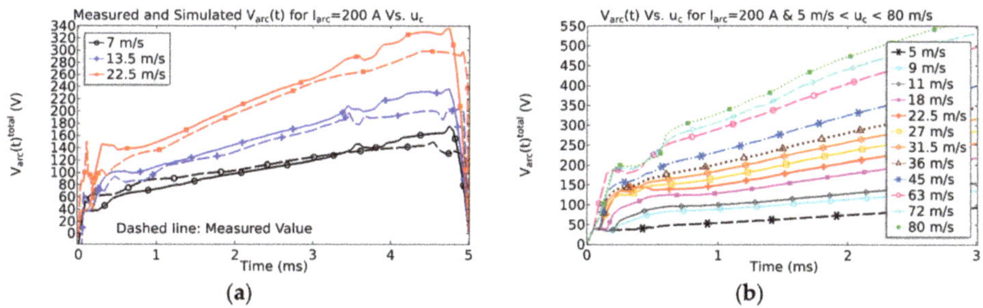

(a) (b)

Figure 16. (**a**) V_{Meas} and $V_{simulated}$ for u_c = 7, 9, 13.5, and 22.5 m/s, (**b**) $V_{simulated}$ in first 1.5 ms for u_c = 5–80 m/s, for 200 A_{peak} arcs.

The failure of fast switches results in current commutation failure, and consequently failure in HVDC breakers and FCLs. The relation between the contact velocity and current is a matter of interruption performance and needs the simulation of transient recovery voltage (TRV) slopes. It is already studied through this validated model and is shown in Figure 6 of [94]. The relationship between the contact velocity and Thomson coil current or the variation in the actuator parameter was already published in [13]. Arc behavior in the failure of FS due to high currents, or insufficient u_c

through data-driven decision making (DDDM), is the target of future studies. DDDM is based on actual data rather than intuition or observation alone. The hard truth is that simulation output alone is not enough. So, making organizational decisions for failure Prediction can be utilized through different modeling techniques [95] like Gaussian Process Regression Models [96] in the failure study.

4.3. Effects of Fast Elongation on the Convective Cooling and V_d/mm

Figure 17a shows the maximum convection flux for 200 A arc at u_c of 5, 7, 9, 13.5, and 27 m/s. It is also evident that convective cooling is related to u_c. Besides, the maximum of the convection is dependent on the maximum of the U (U_{max}), and the U_{max} is dependent on the u_c again [97]. Figure 17b shows the voltage per length for the 200A arc for u_c of 5, 7, 9, 13.5, and 27 m/s within the first 3 ms. Dividing 50 V reported simulated voltages [98] to 8 mm distance between parallel rails [99] and it is already known that elongation increases the cooling and the $V_{arc,\ total}$, but it decreases the arc cross-section. All these changes shall increase the resistance per unit length but, according to Figure 14g, the resistance remains almost fixed and therefore the V_d per unit length inside the arc is decreased.

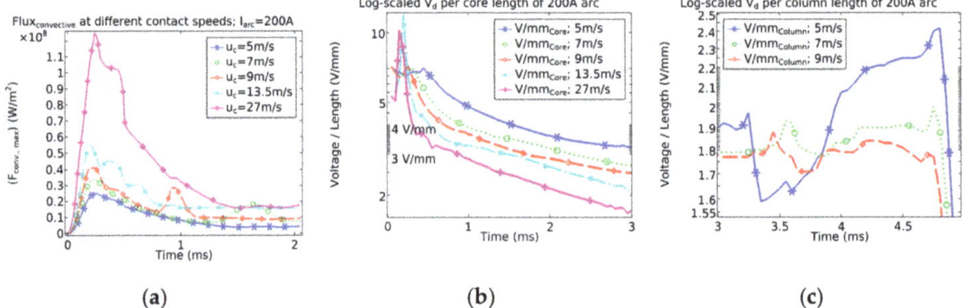

Figure 17. (**a**) Max. convection flux for 200 A arc at u_c = 5, 7, 9, 13.5, 27 m/s. Voltage per length of arc for (**b**) the first 3 ms for 200 A arcs and u_c of 5, 7, 9, 13.5 and 27 m/s, and (**c**) the last 2 ms for 200 A arcs and u_c = 5, 7, 9 m/s.

The physical reason behind this observation is that the elongation makes the arc narrower, so in a fixed length, the plasma volume is reduced, and therefore, lower energy is needed to keep the plasma in the previous condition considering fixed energy per volume for the plasma. As the current is supplied through the current source, the voltage per length is reduced. Except for the first 1 ms in the rest of this period, the voltage per length of arc remains between 2.5–4.5 V/m, which is in line with other measurements [37].

Figure 17c shows the voltage per length of the arc in the last 2 ms before CZ for 200 A arcs at u_c of 5, 7, and 9 m/s. In this period, the voltage per length of arc remains between 1.7–2.3 V/m. The physical reason is that convective cooling is reduced, as is shown in Figure 16b. Therefore, lower input energy is needed to keep the plasma in the previous condition as the loss is reduced. Thus, the voltage per length is reduced again.

5. Conclusions

The concrete findings of this study are pinpointed here:

- Utilizing moving mesh, a detailed 2-D FEM model for an air arc plasma in a prototype FS was presented. The numerical model was established based on MHD and NEC and was validated theoretically and experimentally. Comparing with other researches and experiments, all assumptions were explained in depth.
- The presented model can be adapted to a vast range of geometries, contact opening speeds up to 80 m/s, as well as various arc currents.

- The thermodynamics and electrical behavior of FEA, as well as their visual appearance, were extracted through post-processing of simulated variables and were validated through physical measurements.
- The change of arc mode from constricted to dispersed in FEA was investigated, and the thermo-emission effects on the arc parameters were observed in detail. It was revealed that arc images, arc temperature, and voltage simulated values, as well as the arc behavior, all confirm the change in arc mode.
- The impact of thermionic emission on the contact temperature and the change in σ near the sheath layer, as well as differences between stationary arc and FEA, were explained, and the influence of u_c on the increment in V_{arc} was investigated.
- The effects of fast elongation on the convective cooling and the change of the electric field inside the arc at different periods were extracted.

Author Contributions: Conceptualization, A.K. and K.N.; methodology, software, validation, formal analysis, investigation, resources, data curation, and writing—original draft preparation, A.K.; writing—review and editing, A.K. and K.N.; visualization, A.K.; supervision, K.N.; project administration, A.K.; funding acquisition, K.N. All authors have read and agreed to the published version of the manuscript.

Funding: This research was funded by the Norwegian Research Council under the grant number 280539 and The APC was funded by the Norwegian University of Science and Technology, (NTNU).

Conflicts of Interest: The authors declare no conflict of interest.

References

1. EPRI. *Survey of Fault Current Limiter (FCL) Technologies-Update*; Center of Advanced Power Systems (CAPS): Tallahassee, FL, USA, 2008; p. 54.
2. CIGRE. *Fault Current Limiters in Electrical Medium and High Voltage Systems*; CIGRE: Paris, France, 2003.
3. Steurer, M.; Fröhlich, K.; Holaus, W.; Kaltenegger, K. A novel hybrid current-limiting circuit breaker for medium voltage: Principle and test results. *IEEE Trans. Power Deliv.* **2003**, *18*, 460–467. [CrossRef]
4. Bissal, A. *Modeling and Verification of Ultra-Fast Electro-Mechanical Actuators for HVDC Breakers*; KTH: Stockholm, Sweden, 2015.
5. Park, S.H.; Jang, H.J.; Chong, J.K.; Lee, W.Y. Dynamic analysis of Thomson coil actuator for fast switch of HVDC circuit breaker. In Proceedings of the 3rd International Conference on Electric Power Equipment-Switching Technology (ICEPE-ST), Busan, Korea, 25–28 October 2015; pp. 425–430.
6. Skarby, P.; Steiger, U. An Ultra-fast Disconnecting Switch for a Hybrid HVDC Breaker–a technical breakthrough. In Proceedings of the 2013 CIGRÉ Canada Conference, Calgary, AB, Canada, 9–11 September 2013.
7. Basu, S.; Srivastava, K.D. Analysis of a Fast Acting Circuit Breaker Mechanism Part I: Electrical Aspects. *IEEE Trans. Power Appar. Syst.* **1972**, *PAS-91*, 1197–1203. [CrossRef]
8. Fröhlich, K.; Holaus, W.; Kaltenegger, K.; Steurer, M. High-Speed Current-Limiting Switch. U.S. Patent 6,535,366, 18 March 2003.
9. Holaus, W.; Sartori, S.; Steurer, M.; Fröhlich, K.; Kaltenegger, K. High-Speed Mechanical Switching Point. U.S. Patent 6,636,134, 21 October 2003.
10. Jovcic, D. Fast Commutation of DC Current into a Capacitor Using Moving Contacts. *IEEE Trans. Power Deliv.* **2019**. [CrossRef]
11. Holaus, W.; Fröhlich, K. Ultra-fast switches—A new element for medium voltage fault current limiting switchgear. In Proceedings of the IEEE Power Engineering Society Winter Meeting, New York, NY, USA, 27–31 January 2002; Volume 291, pp. 299–304.
12. Holaus, W. *Ultra Fast Switches–Basic Elements for Future Medium Voltage Switchgear*; Swiss Federal Institute of Technology (ETH): Zurich, Switzerland, 2001.
13. Kadivar, A. Electromagnetic Actuators for Ultra-fast Air Switches to Increase Arc Voltage by Increasing Contact Speed. In Proceedings of the 34th International Power System Conference (PSC-2019), Tehran, Iran, 9–11 December 2019.

14. Wu, Y.; Li, M.; Rong, M.; Wu, Y.; Yang, F. A new model for Thomson-type actuator including the pressure buffer. *Adv. Mech. Eng.* **2015**, *7*. [CrossRef]
15. Vilchis-Rodriguez, D.S.; Shuttleworth, R.; Barnes, M. Double-sided Thomson coil based actuator: Finite element design and performance analysis. In Proceedings of the 8th IET International Conference on Power Electronics, Machines and Drives (PEMD 2016), Glasgow, UK, 19–21 April 2016; pp. 1–6.
16. Bissal, A.; Magnusson, J.; Engdahl, G. Electric to Mechanical Energy Conversion of Linear Ultrafast Electromechanical Actuators Based on Stroke Requirements. *IEEE Trans. Ind. Appl.* **2015**, *51*, 3059–3067. [CrossRef]
17. Peng, C.; Husain, I.; Huang, A.Q.; Lequesne, B.; Briggs, R. A Fast Mechanical Switch for Medium-Voltage Hybrid DC and AC Circuit Breakers. *IEEE Trans. Ind. Appl.* **2016**, *52*, 2911–2918. [CrossRef]
18. Yuan, Z.; He, J.; Pan, Y.; Jing, X.; Zhong, C.; Zhang, N.; Wei, X.; Tang, G. Research on ultra-fast vacuum mechanical switch driven by repulsive force actuator. *Rev. Sci. Instrum.* **2016**, *87*, 125103. [CrossRef]
19. Pei, X.; Smith, A.C.; Shuttleworth, R.; Vilchis-Rodriguez, D.S.; Barnes, M. Fast Operating Moving Coil Actuator for a Vacuum Interrupter. *IEEE Trans. Energy Convers.* **2017**, *32*, 931–940. [CrossRef]
20. Peng, C.; Song, X.; Huang, A.; Husain, I. A Medium-Voltage Hybrid DC Circuit Breaker-Part II: Ultrafast Mechanical Switch. *IEEE J. Emerg. Sel. Top. Power Electron.* **2017**, *5*, 8. [CrossRef]
21. Bissal, A.; Salinas, E. Thomson Coil Based Actuator. U.S. Patent 9,911,562, 6 March 2018.
22. Berger, S. Mathematical approach to model rapidly elongated free-burning arcs in air in electric power circuits. In Proceedings of the 23rd International Conference on Electrical Contacts, Sendai, Japan, 6–9 June 2006; pp. 516–522.
23. Sawicki, A.; Haltof, M. Spectral and integral methods of determining parameters in selected electric arc models with a forced sinusoid current circuit. *Przegląd Elektrotechniczny* **2016**, *65*, 17. [CrossRef]
24. Kadivar, A.; Niayesh, K. Practical methods for electrical and mechanical measurement of high speed elongated arc parameters. *Measurement* **2014**, *55*, 14. [CrossRef]
25. Wu, Z.; Wu, G.; Dapino, M.; Pan, L.; Ni, K. Model for Variable-Length Electrical Arc Plasmas Under AC Conditions. *IEEE Trans. Plasma Sci.* **2015**, *43*, 2730–2737. [CrossRef]
26. Sawicki, A. Problems of modeling an electrical arc with variable geometric dimensions. *Przegląd Elektrotechniczny* **2013**, *89*, 6.
27. Nottingham, W.B. A New Equation for the Static Characteristic of the Normal Electric Arc. *J. Am. Inst. Electr. Eng.* **1923**, *42*, 12–19. [CrossRef]
28. Sawicki, A.; Haltof, M. Mathematical models of electric arc with variable plasma column length used for simulations of processes in gliding arc plasmatrons. *Przegląd Elektrotechniczny* **2016**, *92*, 4. [CrossRef]
29. Li, P.; Zhang, Y.M. Robust sensing of arc length. *IEEE Trans. Instrum. Meas.* **2001**, *50*, 697–704. [CrossRef]
30. Hutchison, R.M. Extraction of Arc Length from Voltage and Current Feedback. U.S. Patent 9,539,662, 10 January 2017.
31. Seeger, M.; Naidis, G.; Steffens, A.; Nordborg, H.; Claessens, M. Investigation of the dielectric recovery in synthetic air in a high voltage circuit breaker. *J. Phys. D Appl. Phys.* **2005**, *38*, 1795.
32. Nakano, T.; Tanaka, Y.; Murai, K.; Uesugi, Y.; Ishijima, T.; Tomita, K.; Suzuki, K.; Shinkai, T. Thermal re-ignition processes of switching arcs with various gas-blast using voltage application highly controlled by powersemiconductors. *J. Phys. D Appl. Phys.* **2018**, *51*, 215202.
33. Peelo, D.F. *Current Interruption Using High Voltage Air-Break Disconnectors*; Technische Universiteit Eindhoven: Eindhoven, The Netherlands, 2004.
34. Li, X.; Wang, X.; Yin, N.; Gao, Q.; Miao, S.; Shan, C.; Huang, X. Simulation on failure analysis of vacuum circuit breaker permanent magnet operating mechanism based on three-parameter method. In Proceedings of the IEEE International Conference on Power System Technology (POWERCON), Wollongong, Australia, 28 September–1 October 2016; pp. 1–6.
35. Jonsson, E.; Aanensen, N.S.; Runde, M. Current Interruption in Air for a Medium-Voltage Load Break Switch. *IEEE Trans. Power Deliv.* **2014**, *29*, 870–875. [CrossRef]
36. Aanensen, N.S.; Jonsson, E.; Runde, M. Air-Flow Investigation for a Medium-Voltage Load Break Switch. *IEEE Trans. Power Deliv.* **2015**, *30*, 299–306. [CrossRef]
37. Støa-Aanensen, N.; Runde, M.; Teigset, A.D. Arcing voltage for a medium-voltage air load break switch. In Proceedings of the 61st IEEE Holm Conference on Electrical Contacts, San Diego, CA, USA, 11–14 October 2015; pp. 101–106.

38. Balestrero, A.; Ghezzi, L.; Popov, M.; Sluis, L.V.D. Current Interruption in Low-Voltage Circuit Breakers. *IEEE Trans. Power Deliv.* **2010**, *25*, 206–211. [CrossRef]
39. Støa-Aanensen, N.; Runde, M.; Jonsson, E.; Teigset, A.D. Empirical Relationships Between Air-Load Break Switch Parameters and Interrupting Performance. *IEEE Trans. Power Deliv.* **2016**, *31*, 278–285. [CrossRef]
40. Sarrailh, P.; Garrigues, L.; Hagelaar, G.J.M.; Boeuf, J.; Sandolache, G.; Rowe, S.W.; Jusselin, B. Two-Dimensional Simulation of the Post-Arc Phase of a Vacuum Circuit Breaker. *IEEE Trans. Plasma Sci.* **2008**, *36*, 1046–1047. [CrossRef]
41. Huber, E.F.J.; Weltmann, K.D.; Froehlich, K. Influence of interrupted current amplitude on the post-arc current and gap recovery after current zero-experiment and simulation. *IEEE Trans. Plasma Sci.* **1999**, *27*, 930–937. [CrossRef]
42. Davidson, P.A. *Introduction to magnetohydrodynamics*, 2nd ed.; TJ International Ltd.: Padstow, UK, 2017.
43. Freton, P.; Gonzalez, J.J.; Gleizes, A. Comparison between a two- and a three-dimensional arc plasma configuration. *J. Phys. D Appl. Phys.* **2000**, *33*, 2442. [CrossRef]
44. Schneidenbach, H.; Uhrlandt, D.; Frank, S.; Seeger, M. Temperature profiles of an ablation-controlled arc in PTFE: II. Simulation of side-on radiances. *J. Phys. D Appl. Phys.* **2007**, *40*, 7402. [CrossRef]
45. Cressault, Y.; Gleizes, A.; Riquel, G. Properties of air-aluminum thermal plasmas. *J. Phys. D Appl. Phys.* **2012**, *45*, 265202. [CrossRef]
46. Vacquie, S. Influence of metal vapours on arc properties. *Pure Appl. Chem.* **1996**, *68*, 1133–1136. [CrossRef]
47. Lago, F.; Gonzalez, J.J.; Freton, P.; Gleizes, A. A numerical modelling of an electric arc and its interaction with the anode: Part I. The two-dimensional model. *J. Phys. D Appl. Phys.* **2004**, *37*, 883. [CrossRef]
48. Rong, M.; Ma, Q.; Wu, Y.; Xu, T.; Murphy, A.B. The influence of electrode erosion on the air arc in a low-voltage circuit breaker. *J. Appl. Phys.* **2009**, *106*, 10. [CrossRef]
49. Jin Ling, Z.; Jiu Dun, Y.; Fang, M.T.C. Electrode evaporation and its effects on thermal arc behavior. *IEEE Trans. Plasma Sci.* **2004**, *32*, 1352–1361. [CrossRef]
50. Murphy, A.B. Influence of metal vapour on arc temperatures in gas–metal arc welding: Convection versus radiation. *J. Phys. D Appl. Phys.* **2013**, *46*, 224004. [CrossRef]
51. Kadivar, A.; Niayesh, K. Two-way Interaction between switching arc and solid surfaces: Distribution of ablated contact and nozzle materials. *J. Phys. D Appl. Phys.* **2019**, 52. [CrossRef]
52. Aubrecht, V.; Bartlova, M.; Coufal, O. Radiative emission from air thermal plasmas with vapour of Cu or W. *J. Phys. D Appl. Phys.* **2010**, *43*, 434007. [CrossRef]
53. Capitelli, M.; Colonna, G.; Gorse, C.; D'Angola, A. Transport properties of high temperature air in local thermodynamic equilibrium. *Eur. Phys. J. D* **2000**, *11*, 11. [CrossRef]
54. D'Angola, A.; Colonna, G.; Gorse, C.; Capitelli, M. Thermodynamic and transport properties in equilibrium air plasmas in a wide pressure and temperature range. *Eur. Phys. J. D* **2008**, *46*, 22. [CrossRef]
55. Dixon, C.M.; Yan, J.D.; Fang, M.T.C. A comparison of three radiation models for the calculation of nozzle arcs. *J. Phys. D Appl. Phys.* **2004**, *37*, 3309. [CrossRef]
56. Reichert, F.; Rümpler, C.; Berger, F. Application of different radiation models in the simulation of air plasma flows. In Proceedings of the 17th International Conference on Gas Discharges and Their Applications, Cardiff, UK, 7–12 September 2008; pp. 141–144.
57. Sylvain, C.; Jean-Luc, M. Theoretical prediction of non-thermionic arc cathode erosion rate including both vaporization and melting of the surface. *Plasma Sources Sci. Technol.* **2000**, *9*, 239. [CrossRef]
58. Naghizadeh-Kashani, Y.; Cressault, Y.; Gleizes, A. Net emission coefficient of air thermal plasmas. *J. Phys. D Appl. Phys.* **2002**, *35*, 2925. [CrossRef]
59. Capitelli, M.; Colonna, G.; D'Angola, A. *Fundamental Aspects of Plasma Chemical Physics, Thermodynamics*, 1st ed.; Springer: New York, NY, USA, 2012; Volume 1, p. 310.
60. Capitelli, M.; Bruno, D.; Laricchiuta, A. *Fundamental Aspects of Plasma Chemical Physics, Transport*, 1st ed.; Springer: New York, NY, USA, 2013; p. 352. [CrossRef]
61. Koelman, P.M.J.; Mousavi, S.T.; Perillo, R.; Graef, W.; Mihailova, D.B.; van Dijk, J. Studying complex chemistries using PLASIMO's global model. *J. Phys. Conf. Ser.* **2016**, *682*, 012034. [CrossRef]
62. Kadivar, A.; Niayesh, K. Metal vapor content of an electric arc initiated by exploding wire in a model N2 circuit breaker: Simulation and experiment. *J. Phys. D Appl. Phys.* **2020**, in press.
63. Nave, C.L. Magnetic Properties of Solids. Available online: http://hyperphysics.phy-astr.gsu.edu/hbase/Tables/magprop.html (accessed on 18 August 2018).

64. Artale, C.; Fermepin, S.; Forti, M.; Latino, M.; Quintero, M.; Granja, L.; Sacanell, J.; Polla, G.; Levy, P. Electric and magnetic properties of PMMA/manganite composites. *Phys. B Condens. Matter* **2009**, *404*, 2760–2762. [CrossRef]
65. Munoz, J.; Rojo, M.; Parrefio, A.; Margineda, J. Automatic measurement of permittivity and permeability at microwave frequencies using normal and oblique free-wave incidence with focused beam. *IEEE Trans. Instrum. Meas.* **1998**, *47*, 886–892. [CrossRef]
66. Haynes, W.M.; Lide, D.R.; Bruno, T.J. *CRC Handbook of Chemistry and Physics*, 96th ed.; CRC Press: Boca Raton, FL, USA, 2015; Volume 1.
67. Dellino, G.; Meloni, C. *Uncertainty Management in Simulation-Optimization of Complex Systems, Algorithms and Applications*; Springer: Boston, MA, USA, 2015; Volume 59.
68. Kriegel, M.; Uzelac, N. Simulations as Verification Tool for Design and Performance Evaluation of Switchgears. In *Switching Equipment*; Ito, H., Ed.; Springer International Publishing: Cham, Switzerland, 2019; pp. 379–397. [CrossRef]
69. Dijk, J.V.; Kroesen, G.M.W.; Bogaerts, A. Plasma modelling and numerical simulation. *J. Phys. D Appl. Phys.* **2009**, *42*, 190301. [CrossRef]
70. Crowell, C.R. The Richardson constant for thermionic emission in Schottky barrier diodes. *Solid State Electron.* **1965**, *8*, 395–399. [CrossRef]
71. Hauser, A.; Branston, D.W. Numerical simulation of a moving arc in 3D. In Proceedings of the 17th International Conference on Gas Discharges and Their Applications, Cardiff, UK, 7–12 September 2008; pp. 213–216.
72. Proskurovsky, D.I. Explosive Electron Emission from Liquid-Metal Cathodes. *IEEE Trans. Plasma Sci.* **2009**, *37*, 1348–1361. [CrossRef]
73. Wilson, R.G. Vacuum Thermionic Work Functions of Polycrystalline Nb, Mo, Ta, W, Re, Os, and Ir. *J. Appl. Phys.* **1966**, *37*, 3170–3172. [CrossRef]
74. Modinos, A. Theory of thermionic emission. *Surface Sci.* **1982**, *115*, 469–500. [CrossRef]
75. Cayla, F.; Freton, P.; Gonzalez, J.J. Arc/Cathode Interaction Model. *IEEE Trans. Plasma Sci.* **2008**, *36*, 1944–1954. [CrossRef]
76. Mutzke, A.; Rüther, T.; Lindmayer, M.; Kurrat, M. Arc behavior in low-voltage arc chambers. *Eur. Phys. J. Appl. Phys.* **2010**, *49*, 22910. [CrossRef]
77. Barral, N.; Alauzet, F. Three-dimensional CFD simulations with large displacement of the geometries using a connectivity-change moving mesh approach. *Eng. Comput.* **2019**, *35*, 397–422. [CrossRef]
78. COMSOL. *Version 5.4a, Reference Manual*; COMSOL Multiphysics: Stockholm, Sweden, 2019.
79. Schenk, O.; Gärtner, K.; Fichtner, W.; Stricker, A. PARDISO: A high-performance serial and parallel sparse linear solver in semiconductor device simulation. *Future Gener. Comput. Syst.* **2001**, *18*, 69–78. [CrossRef]
80. Huguenot, P.; Kumar, H.; Wheatley, V.; Jeltsch, R.; Schwab, C. Numerical Simulations of High Current Arc in CB. In Proceedings of the 24th International Conference on Electrical Contacts, Saint-Malo, France, 9–12 June 2008.
81. Schlichting, H.; Gersten, K. *Boundary-Layer Theory*, 9th ed.; Springer: Berlin Heidelberg, Germany, 2017. [CrossRef]
82. Krouchinin, A.M.; Sawicki, A.; Częstochowska, P. *A Theory of Electrical Arc Heating*; Technical University of Częstochowa: Częstochowa, Poland, 2003. [CrossRef]
83. Mutzke, A.; Ruther, T.; Kurrat, M.; Lindmayer, M.; Wilkening, E.D. Modeling the Arc Splitting Process in Low-Voltage Arc Chutes. In Proceedings of the 53rd IEEE Holm conference on Electrical Contacts, Pittsburgh, PA, USA, 16–19 September 2007; pp. 175–182.
84. Howatson, A.M. *An Introduction to Gas Discharges*, 2nd ed.; Pergamon Press: Oxford, NY, USA, 1976.
85. Jiao-Min, L.; Jie, Z.; Bai-Hong, T.; Zhen-Zhou, W. A Study On Temperature Distribution of arc Cross Section of Low-Voltage Apparatus. In Proceedings of the 2005 International Conference on Machine Learning and Cybernetics, Guangzhou, China, 18–21 August 2005; pp. 315–320.
86. Valerian, A.N. Anode layer in a high-current arc in atmospheric pressure nitrogen. *J. Phys. D Appl. Phys.* **2005**, *38*, 4082. [CrossRef]
87. Niayesh, K.; Runde, M. *Power Switching Components, Theory, Applications and Future Trends*, 1st ed.; Springer: New York, NY, USA, 2017; p. 249. [CrossRef]

88. Benilov, M.S. Analysis of ionization non-equilibrium in the near-cathode region of atmospheric-pressure arcs. *J. Phys. D Appl. Phys.* **1999**, *32*, 257. [CrossRef]
89. Rei, H.; Yasunobu, Y.; Toshiro, M. Anode-fall and cathode-fall voltages of air arc in atmosphere between silver electrodes. *J. Phys. D Appl. Phys.* **2003**, *36*, 1097. [CrossRef]
90. McBride, J.W.; Weaver, P.M. Review of arcing phenomena in low voltage current limiting circuit breakers. *Sci. Meas. Tech. IEE Proc.* **2001**, *148*, 1–7. [CrossRef]
91. Watanabe, S.; Kokura, K.; Minoda, K.; Sato, S. Characteristics of the Arc Voltage of a High-Current Air Arc in a Sealed Chamber. *Electr. Eng.* **2014**, *186*, 9. [CrossRef]
92. Rong, M.; Yang, F.; Wu, Y.; Murphy, A.B.; Wang, W.; Guo, J. Simulation of Arc Characteristics in Miniature Circuit Breaker. *IEEE Trans. Plasma Sci.* **2010**, *38*, 2306–2311. [CrossRef]
93. Kadivar, A.; Niayesh, K. Simulation of free burning arcs and arcs Inside Cylindrical Tubes initiated by exploding wires. In Proceedings of the 5th International Conference on Electric Power Equipment-Switching Technology (ICEPE-ST), Kitakyushu, Japan, 13–16 October 2019.
94. Kadivar, A.; Niayesh, K. Dielectric Recovery of Ultrafast-Commutating Switches used for HVDC and Fault Current Limiting Applications. In Proceedings of the 5th International Conference on Electric Power Equipment-Switching Technology (ICEPE-ST), Kitakyushu, Japan, 13–16 October 2019.
95. Liu, K.; Ashwin, T.R.; Hu, X.; Lucu, M.; Widanage, W.D. An evaluation study of different modelling techniques for calendar ageing prediction of lithium-ion batteries. *Renew. Sustain. Energy Rev.* **2020**, *131*, 110017. [CrossRef]
96. Liu, K.; Hu, X.; Wei, Z.; Li, Y.; Jiang, Y. Modified Gaussian Process Regression Models for Cyclic Capacity Prediction of Lithium-Ion Batteries. *IEEE Trans. Transp. Electrif.* **2019**, *5*, 1225–1236. [CrossRef]
97. Jong-Chul, L.; Kim, Y.J. Calculation of the interruption Process of a self-blast circuit breaker. *IEEE Trans. Magn.* **2005**, *41*, 1592–1595. [CrossRef]
98. Barbu, B.; Berger, F. Sheath layer modeling for switching arcs. In Proceedings of the ICEC 2014 the 27th International Conference on Electrical Contacts, Dresden, Germany, 22–26 June 2014; pp. 1–6.
99. González, D.; Berger, F. Arc running behaviour between parallel rails of different metals and compounds used in miniature circuit breakers. In Proceedings of the 2013 48th International Universities' Power Engineering Conference (UPEC), Dublin, Ireland, 2–5 September 2013; pp. 1–6.

Article

Properties of Vacuum Arcs Generated by Switching RMF Contacts at Different Ignition Positions

Sergey Gortschakow [1,*], Steffen Franke [1], Ralf Methling [1], Diego Gonzalez [1], Andreas Lawall [2], Erik D. Taylor [2] and Frank Graskowski [2]

1 Department of Plasma Radiation Techniques, Leibniz Institute for Plasma Science and Technology, 17489 Greifswald, Germany; steffen.franke@inp-greifswald.de (S.F.); methling@inp-greifswald.de (R.M.); diego.gonzalez@inp-greifswald.de (D.G.)

2 Siemens AG, 13629 Berlin, Germany; andreas.lawall@siemens.com (A.L.); erik-d.taylor@siemens.com (E.D.T.); frank.graskowski@siemens.com (F.G.)

* Correspondence: sergey.gortschakow@inp-greifswald.de

Received: 17 September 2020; Accepted: 21 October 2020; Published: 26 October 2020

Abstract: The influence of initiation behavior of the drawn arc on the arc motion, on arc characteristics during the active phase, as well as on the post-arc parameters, was studied. The study was focused on arc dynamics, determination of the anode surface temperature after current interruption, and diagnostics of metal vapor density after current zero crossing. Different optical diagnostics, namely high-speed camera video enhanced by narrow-band optical filters, near infrared spectroscopy, and optical absorption spectroscopy was applied. The initiation behavior of the drawn arc had a clear influence on arc parameters. Higher local electrode temperature occurs in case of the electrodes with ignition point near the outer electrode boundary. This further causes an enhanced density of chromium vapor, even in cases with lower arc duration. The results of this study are important for design development of switching RMF contacts for future green energy applications.

Keywords: vacuum interrupter; switching arc; optical diagnostics

---

1. Introduction

Vacuum technology provides environmentally compatible and emission-free solutions for switching applications in power grids [1,2]. Some advantages of vacuum circuit breakers include a high number of operations under standard load conditions, safe and reproducible short-circuit current interruption capability, and maintenance-free operation. Further development of such devices requires fundamental knowledge about the switching process and interaction of materials used with the working medium—the vacuum arc plasma.

The vacuum arc is ignited by the electrode separation under the current load. While the arc appears as diffuse glow for currents typically below 10 kA [1–3], constriction effects dominate at higher currents. The disadvantage of a constricted arc is that it causes a localized thermal load on the electrode, which leads to enhanced electrode erosion due to melting and evaporation. Therefore, various measures are applied in switching devices for arc control. The application of external magnetic fields helps either to hold the arc in diffuse stage (axial magnetic field (AMF)) or causes the rotation of the constricted arc (radial magnetic field (RMF)) or transverse magnetic field (TMF)) [1].

The RMF/TMF contacts have been in use for many decades. In the pioneering works [4–6], the two different contact shapes were suggested. Corresponding principles are used in the modern contact systems in vacuum interrupters and give inspiration for improved contact geometries [7]. The choice of contact design depends on the electrical and thermal load in a certain application.

In case of successful current interruption, the arc is terminated immediately after the current zero (CZ) crossing. However, there are various factors that could prevent the current interruption. When the density of neutral metal vapor is high enough, ionization due to appearance of transient recovery voltage can lead to arc reignition and disconnection failure [8]. The anode surface is the main source of neutral vapor. The anode activity is particularly high if high-current anode modes occur [9]. In case of RMF contacts, the arc movement prevents formation of stationary anode spots. However, the arc needs certain time to maintain the magnetic field for the start of arc rotation. During this stage, the arc position is fixed, which leads to a localized electrode heating. In later stages, the same position could be repeatedly heated, due to a high rotation speed. Since thermal dissipation is a relatively slow process, the position of arc initiation can potentially reach higher temperatures compared to the surrounding material and thereby influence the performance of the contacts. Thus, studying the influence of arc ignition position on the arc dynamics, spatio-temporal evolutions of electrode surface temperature, and vapor density after current interruption is of great practical importance for the understanding of switching behavior and development/improvement of the contact design.

Optical diagnostics offer numerous methods for the characterization of the arc plasma and electrodes. High-speed camera techniques are widely used for observation of arc dynamics, such as appearance of certain high-current modes, mode transitions, arc constriction, rotation speed, etc., [10–13]. Various high-speed cameras are used for acquisition of temporal evolution of light emitted by the arc (mainly in visible range). Typically, an acquisition frequency of 5 k–100 k fps is applied.

Optical emission spectroscopy (OES) can determine the surface temperature. A big challenge for surface temperature measurements is the presence of the arc in front of the electrode. The arc radiation is quite strong during the whole arcing phase and disappears immediately after current zero crossing. This behavior of plasma radiation gives the opportunity for undisturbed access to the electrode surface after the current zero. The corresponding measurements techniques have been reported in [14–16]. One of the possibilities is to use a compact near infrared spectrometer [14], which is adjusted to the position of interest by an optical system. The method works as long as the surface emits enough thermal radiation.

Evaluation of the ground state density of neutral species (Cu and Cr atoms) can be performed by optical absorption spectroscopy (OAS) [17–21]. The absorption measurement setup typically consists of a light source, optical system for beam adjustment and a spectrograph for registration of absorption spectra. Usually, broadband light sources are applied since they provide enough power in the visible spectral region and, thus, allow for a choice of various wavelength ranges of interest. The measurements are of feasible complexity as long as they are performed after the current zero crossing, when the emission from arc plasma is negligibly small [17–20]. Investigations in the active phase, however, require much more effort, e.g., arrangement of a second optical path for registration of plasma radiation from the probed region only [21]. This technique is very complicated and works well in the case of non-moving arc only. Therefore, a setup with single optical path was used for measurements after current interruption.

In the present study, the influence of the arc ignition position on local surface temperature of the anode and vapor density after current zero crossing was investigated in a model vacuum interrupter using the above-mentioned optical diagnostics under realistic operation conditions.

2. Materials and Methods

2.1. Experimental Setup

Figure 1 presents the experimental setup, comprising a model vacuum interrupter (vacuum chamber 1 with pneumatic drive 2), a high-current generator (visible in the background of Figure 1), as well as electrical and optical diagnostics (3–8).

A vacuum chamber with optical access described in [22] has been used. It contains four optical viewports allowing for simultaneous acquisition of several optical signals. The electrodes are moved by a pneumatic drive with an opening speed of about 1 m/s. The fixed electrode was powered as an anode. The maximum stroke of the drive system was about 10 mm. The electrode distance at which the arc interruption occurs depends on the opening speed and the instant of contact separation start and is typically 3–8 mm.

Figure 1. Setup for studies on vacuum arcs: (**1**) vacuum chamber, (**2**) pumping system, (**3**) high-speed camera (general arc dynamics), (**4**) two high-speed cameras equipped with narrow-band optical filters, (**5**) NIR optics, (**6**) NIR spectrometer, (**7**) Xe flash lamp, (**8**) 0.75 m imaging spectrograph.

2.2. Power Source

Two different power sources have been used. For electrode surface cleaning a pulsed DC current of 400 A and 20 ms duration has been applied [22]. The second power generator, providing 50 Hz equivalent current with a magnitude of about 28 kA (peak value), was used for the main experiments. The arc duration varied between 3 and 8 ms, depending on the time instant of electrode separation.

2.3. Electrodes

Spiral RMF electrodes made of CuCr alloy were used. Electrode diameter was 34 mm. Contact spots at defined positions for arc initialization have been created mechanically. Figure 2 shows examples of prepared electrodes.

These initial contact spots had diameters of about 3 mm. Only the cathodes were modified in order to reduce the influence on the anode as much as possible

Figure 3 demonstrates the working principle of a RMF contact pair. When the electrode separation starts, the last connection point between the electrodes serves as an arc ignition position due to fast local overheating (bridge explosion). The arc column constricts due to the action of self-induced magnetic field. The spiral form of the contact produces an additional magnetic field, which pushes the arc toward the outer electrode boundary. As far as the constricted arc column reaches the outer boundary, the joint action of the current flow (blue tracer in Figure 3) and magnetic field (green arrows in Figure 3) causes a Lorentz force (direction is shown by red arrows in Figure 3), which rotates the arc.

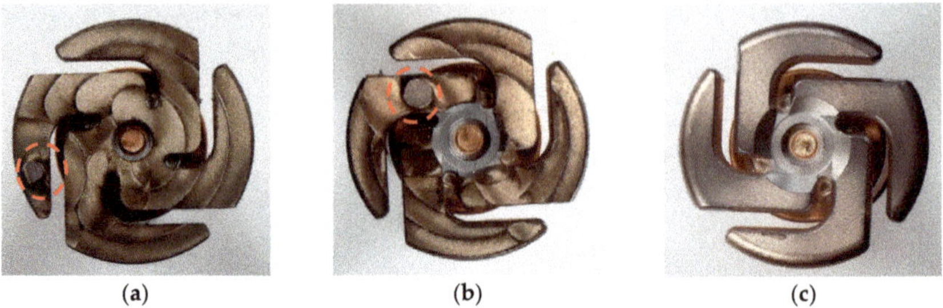

Figure 2. Examples of CuCr electrodes used in the study. Red circles show the prepared contact spots for arc ignition, type A (**a**) and type B (**b**). The anode (**c**) was not modified.

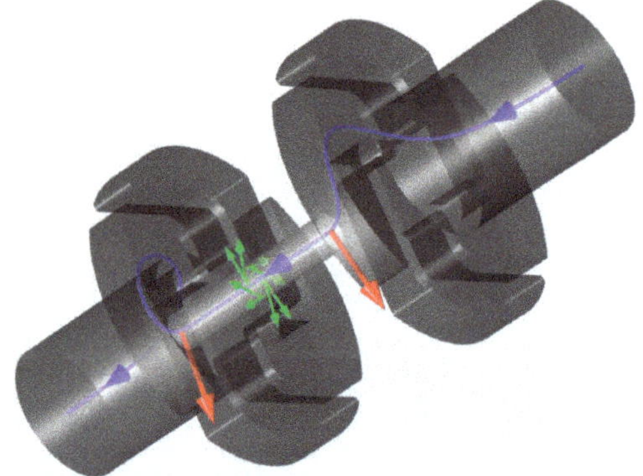

Figure 3. Schematic representation of RMF contact working principle. Blue tracer shows the current flow, green arrows represent the direction of magnetic flux density, and red arrows show the direction of resulting Lorentz force.

2.4. Diagnostics

Electrical and optical diagnostics was used for characterization of the arc behavior. Several high-speed cameras acquired the arc dynamics (Figure 1). Near infrared (NIR) spectroscopy determined the time-resolved anode surface temperature after current interruption. Broadband absorption spectroscopy in turn determined the vapor density with respect to different initial arc positions.

2.4.1. Electrical Diagnostics

Arc current and arc voltage have been measured by corresponding probes. A Rogowski current probe (model CWT 1500, PEM) measured the current evolution. The arc voltage was acquired using a high voltage probe (P6015A, Tektronix). The signals were registered by a transient recorder (GEN7t, HBM).

2.4.2. Arc Images

Several high-speed cameras registered the arc dynamics (Motion Pro Y4, IDT) with a repetition/frame rate of 20,000 fps and an exposure time of 1 µs. Standard camera lenses were used. One camera monitored the ignition position and general arc dynamics. In addition, two cameras equipped with narrow-band metal-interference filters transmitting at 521 nm with FMHW 1 nm (Cu I line) and at 494 nm with FWHM 4 nm (several Cu II lines) separated the dynamics of copper atoms and ions during the arcing.

2.4.3. Determination of Surface Temperature

Near infrared spectroscopic measurements (NIR in Figure 1) have been used for evaluation of the local anode surface temperature after current interruption (after the time instant 10 ms in Figure 4b) at the points of arc ignition. A NIR spectrometer (C1142GA, Hamamatsu) with a spectral range of 900–1650 nm and a temporal resolution of 1.25 ms (exposure time 200 µs) was used for spectral measurements. After the acquisition, the spectra were processed according to the routine presented in [14] to determine the anode surface temperature. Figure 4 summarizes the information about NIR diagnostics. The red spot in Figure 4a shows the region, from which the anode surface radiation was collected. The measurement spot was about 1 mm in diameter and focused on the arc ignition position, i.e., in case of electrodes of type A, the position shown by red circle in Figure 2a was used, in case of the electrodes of type B, the corresponding position shown in Figure 2b. The measurements started around the current maximum with an acquisition time of 200 µs and frequency of one spectrum each 1.25 ms.

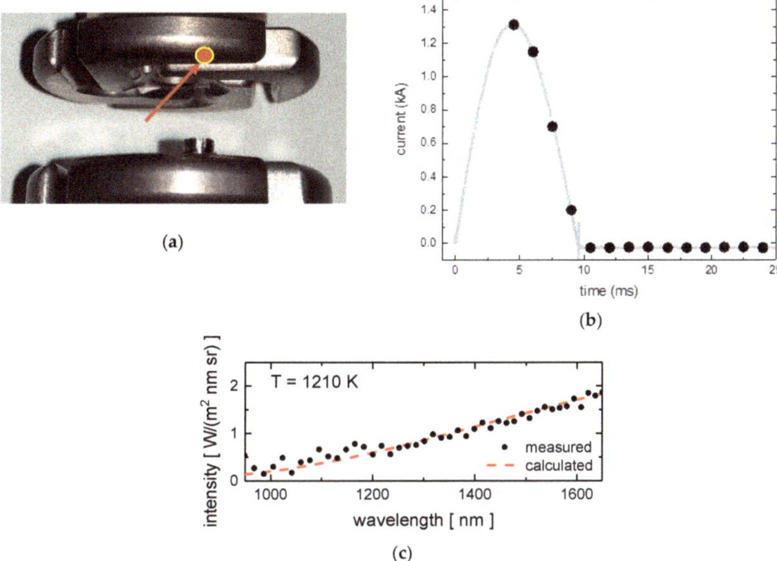

Figure 4. (**a**) Measurement position in case of NIR spectroscopy (red circle). Anode on the top, cathode in the bottom. (**b**) Acquisition instants of NIR spectroscopy related to the current shape. (**c**) Sample of NIR spectrum suitable for temperature evaluation: instant of time 1.25 ms after current interruption.

Figure 4c shows an example of temperature evaluation for 1.25 ms after current interruption (point around 11.25 ms in Figure 4b) for the contact pair of type B. The surface temperature was calculated using diagnostic techniques described in [14]. A tungsten strip lamp was used for determination of wavelength dependent windows transmission, as well as for spectral calibration of NIR spectrometer. The temperature arises from the shape of measured intensity (Figure 4c) by comparison with a Planck radiator with given temperature. Details of the methods are described in [14].

2.4.4. Absorption Spectroscopy for Species Density Determination

Broadband optical absorption spectroscopy (OAS) determined the chromium vapor density after extinction of arc plasma (after current interruption). This technique is based on evaluation of absorption spectra in the wavelength range where the resonance lines of material of interest are present.

A pulsed DC high-intensity Xenon lamp acted as a background radiation source. It emits a Planck-like radiation of 12,000 K with a maximum power of 1 MW [23]. The Xenon lamp is positioned on the right-side window of the vacuum chamber (Figure 1). Its radiation is directed through the electrode system and coupled to the spectrometer entrance slit that is placed at the opposite window. By using a deflecting and a focusing mirror, the electrode gap is observed along a line parallel to the electrode surfaces (dashed line in Figure 5a). The slit width was about 50 µm. The spatial position of spectra acquisition was about 1 mm away from the anode surface. The lamp and the spectrograph start at a desired instant (Figure 5b). The Xe lamp started at about 9.5 ms (green curve in Figure 5b) and reached its maximum intensity close to the instant of acquisition time (grey curve denoted as OAS in Figure 5c). The radiation is spectrally dispersed using a Czerny–Turner type spectrograph (Shamrock 750, Andor Technology Ltd.) with a 0.75 m focal length equipped with an intensified charge coupled device camera (iStar, Andor Technology Ltd.). Absorption spectra were acquired immediately after the current zero crossing with a delay between 100 µs and 300 µs, which was necessary to study the decaying plasma. A spectral interval between 423–431 nm was chosen to study the Cr I density, which contains spectral lines of different resonance transitions of 425.43, 427.78, and 428.97 nm. Figure 4c presents an example of an acquired spectrum for Cr I 428.97 nm line. The density of absorbing species is proportional to the area under the curve. More details about the method and the corresponding theory are described in [20,21].

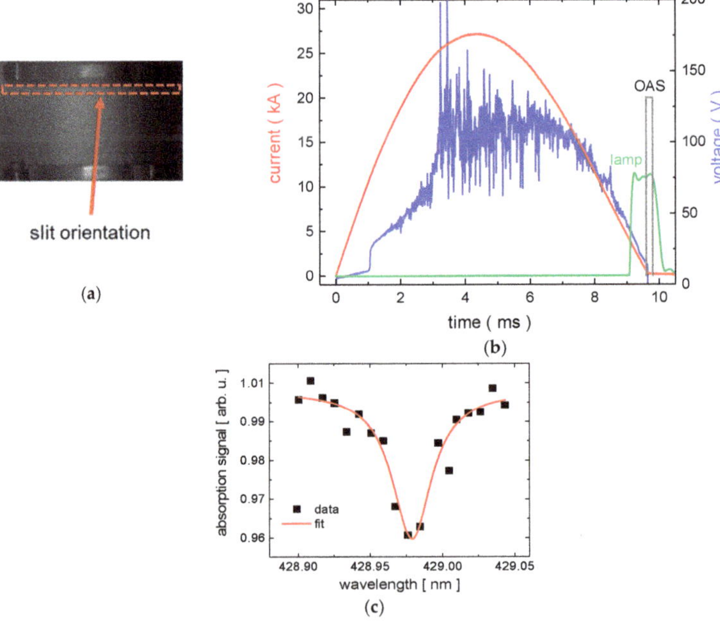

Figure 5. (**a**) Position of spectrograph slit for absorption spectra acquisition. Slit width was 50 µm; distance to the anode was about 1 mm. (**b**) Temporal evolution of current (red) and voltage (blue) along with sequence of control signals during the absorption spectroscopy measurements: driving current of Xe lamp (green) and exposure signal for spectrum acquisition (grey). (**c**) Example of Cr I 428.97 nm line fit procedure.

3. Results and Discussion

3.1. Electrode Preparation

The switching contacts in real vacuum interrupters are manufactured under special conditions in a clean atmosphere [1]. These conditions are typically not available in research facilities. Therefore, some cleaning procedure must be applied prior to a study of arc properties in order to remove adsorbed micro particles, lubricant from the machining processing, dielectric layers, water, etc. The cleaning procedure comprises processing in an ultrasonic bath with distilled water and a degreasing fluid, surface cleaning with isopropanol, and application of low-current arc discharges (<400 A) of less than 30 ms duration for final conditioning of the surface after the mounting inside the vacuum chamber and pumping to ultra-high vacuum. First shots with non-conditioned electrodes show instabilities in the voltage behavior (Figure 6), which is related to stochastic spot formation on the electrode surface at the positions of adsorbed impurities. The spot formation is accompanied by voltage jumps (blue curve in Figure 6) in agreement with different arc voltages for cathode spots I and II [24]. The high-speed videos confirm the unstable arc root positions in this case. After typically 10 shots, the voltage curve becomes much smoother (green curve in Figure 6), indicating the end of the conditioning process.

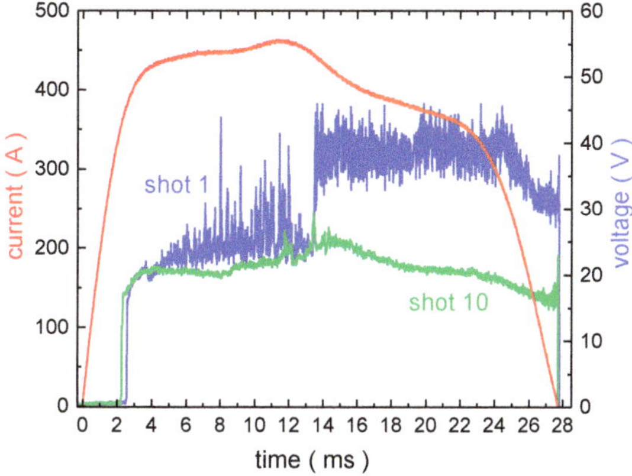

Figure 6. Electrical signals during the conditioning process. Red curve—arc current; blue and green curves—arc voltage during two different shots; blue curve before conditioning and green curve after conditioning.

3.2. Arc Dynamics

Figure 7 presents the typical evolutions of arc voltage and current along the discharge with spectrally filtered images. The ignition instant in the considered example (the start of electrode separation) was about 2.2 ms (Figure 7b,c). The ignition position in present study was defined by mechanical surface modification, as described in Section 2.3. After the ignition, the arc spreads fast over an area, which is larger than the ignition spot and remains diffuse in the present example up to about 3.6 ms (Figure 7d,e). During the diffuse stage, the arc is formed by a nearly homogeneous distributed cathode spots, uniformly glowing anode and a diffuse arc column. The spatial distribution of atoms and ions is very similar at this stage. The bridge explosion gives enough material, which fills a significant volume due to small electrode distance. The arc becomes constricted around the last contact spot (ignition point) (Figure 7f,g) after a certain amount of time, which is necessary to build up enough electromagnetic force (Lorentz force) for constriction process. Notice that even after constriction, the cathode spots cover nearly the whole electrode surface. Their distribution becomes inhomogeneous with higher spot density around the position of constricted arc attachment. Opposite to this, the anode attachment is localized and moves in accordance with actual electromagnetic force direction. Thus, the arc rotation means moving group of cathode spots, opposite anode spot and the arc column, which connects both attachments. The arc trajectory (moving position of highest intensity) is close to the outer electrode boundary. The frames (h–k) show the instants of time when the arc was passing through the ignition position. The spatial distributions of atoms and ions start to differ from each other. At this stage, the atomic radiation is more constricted and has higher intensity near to the cathode surface, while the ionic radiation is distributed wider and more homogeneously over the gap region. This qualitative picture changes as soon as the arc starts to rotate, i.e., after about 3.6 ms in the present example. During rotation, the arc column is dominated by the ions. Their spatial distribution is homogeneous within the emitting area. The atoms follow the ions with some delay. The radiation remains most intense near the cathode surface and is less pronounced in the arc column. The area of atomic radiation near the cathode is much larger compared to that of ionic radiation. The cathode surface is nearly completely melted as will be shown below. Notice, that an arc constriction does not imply that the electrode surface outside of the constriction area is not active anymore. Detailed analysis of high-speed images shows that some weak cathode spots continue to exist on the broader electrode area, and thus represent an intense source of atoms. In addition, the atoms, which are evaporated by melted surface, can become excited due to reaction kinetic processes, like e.g., ion recombination, de-excitation cascades, or charge transfer reactions between copper species. Detailed explanation requires a space- and time-dependent collisional-radiative model, which is missing due to its complexity.

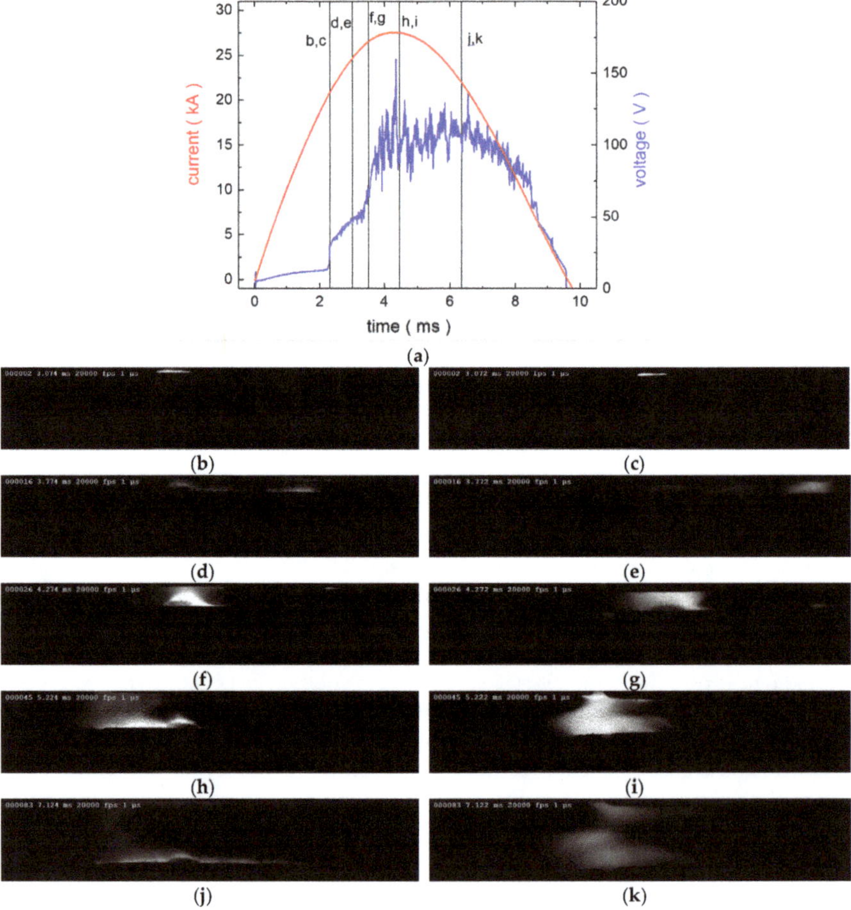

Figure 7. (**a**) Temporal evolution of arc current and voltage. Vertical lines show the instants for which the arc images are presented. (**b–k**) Spectrally filtered arc images at different time instants. Images (**b**,**d**,**f**,**h**,**j**) show Cu I emission, (**c**,**e**,**g**,**i**,**k**) that of Cu II. Electrode type A.

Figure 8 presents the arc rotation frequency along with the arc current shape. It was determined from the arc images by evaluation of the instants at which the arc column passed the same spatial position on the electrode surface. The rotation frequency depends on several factors, like. e.g., arc current value, magnitude of magnetic field flux and the electrode distance. At the beginning, the rotation is slow. The frequency reaches its maximum value, which is about 4 kHz in current example, around the peak current. Then, the frequency follows the current, decreasing up to the instant of about 6.5 ms. The rotation becomes slightly faster for the later time instants, due to accumulated effects of phase shift between the arc current and induced magnetic flux and increasing electrode distance. The initial arc position has no influence on rotation frequency taking into account the accuracy of applied method. In the case of position B (which is closer to the electrode centre), the arc starts to move closer to the electrode axis and reaches the outer rotational position slightly later, comparing to ignition position A. The analysis of the high-speed images clarifies that most of the electrode surface becomes molten (or at least close to the molten state) after several rotation cycles. Thus, for the same arc duration a comparable molten surface area will be reached later for the case with ignition position B.

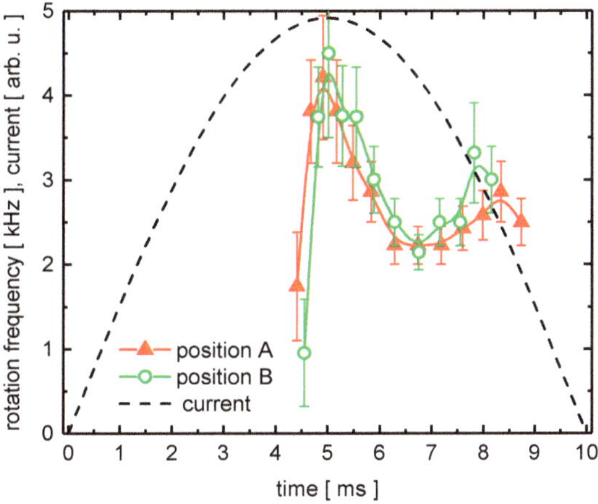

Figure 8. Rotation frequency of the arc determined from high-speed images. Red triangles—contact system of type A; green circles—those of type B. Arc current evolution is presented by dashed curve.

3.3. Anode Surface Temperature

Figure 9 presents the temporal evolutions of local anode surface temperature after the current zero crossing at the position of arc ignition, i.e., at localized spot of about 1 mm in diameter. In case of electrodes of type A, the position shown in Figure 2a was used, in case of the electrodes of type B, the position shown in Figure 2b. Notice that the temperature in the other regions of the anode surface can significantly differ from obtained values. However, it is expected that measured values give an estimate for the maximum anode temperature after current interruption, since the accumulated arc residence time in those region is higher than in surrounding regions. For type A electrodes, the initial temperature tends to increase with the arc duration. Its initial value was around 1300 K. Longer arcing time leads to slower temperature decay. Since the arc is rotating, the measurement point could be slightly cooled down between two subsequent rotations. However, with increasing arc duration the amount of melted material becomes bigger. This is clearly visible in the videos. Consequently, the melted pool needs more time for cooling down. The type B electrodes show less variability in the temperature. The position at which the temperature was measured is in this case closer to the electrode axis. The arc residence time in that region is much shorter comparing to the outer region. Consequently, the temperature is lower, its value is about 1200 K. Slower temperature decay in this region is mainly caused by contact design. The cooling path for the surface is going toward the stem.

The electrode temperature is obviously higher in the outer regions (Figure 9). Thus, those regions act as a heat source for the inner part of the electrode after current interruption. Therefore, a temperature stagnation occurs in those regions.

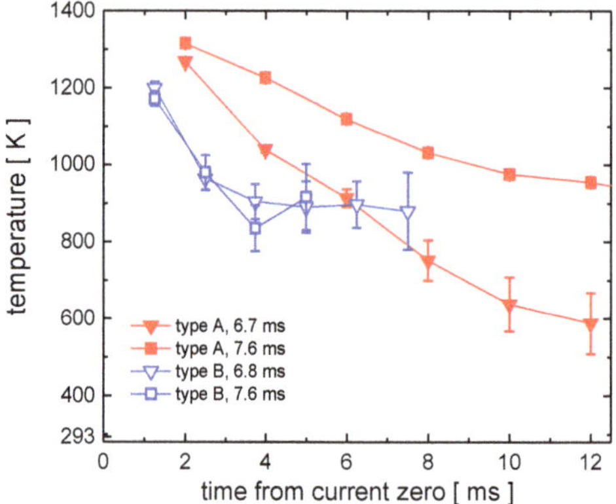

Figure 9. Evolution of surface temperature measured at ignition position after the arc extinction. Arc duration time is indicated. Uncertainty of the numerical method is shown. Total measurements uncertainty is estimated as ±50 K [14].

3.4. Cr Density after Current Interruption

Figure 10 shows the results of absorption measurements for Cr I vapor density after current zero crossing. Due to low spectrally resolved radiation intensity, the acquisition of only one spectrum per shot is possible. Therefore, several measurements (up to five for each contact pair) at the same peak current value of 28 kA have been performed. Due to the presence of many mechanical parts in the driving and in the switching systems it is not possible to get exactly the same arc duration in different shots and the same instant of absorption spectrum acquisition. On the other hand, the reproducibility of plasma characteristics seems to be quite stable, when the same arc duration occurs. Thus, different shots give an estimation of temporal evolution after current zero crossing. The temporal region within first 500 μs is of special importance for interruption process. During this time period, the fast increasing transient recovery voltage can force a breakdown if the species densities are high enough. Atomic species do not follow the electric field and have enough residence time between the electrodes to get ionized. The density of neutral species has the highest value direct after the current zero crossing instant and shows a nearly linear decay within the first microseconds after current interruption [20,21].

The arc duration has an insignificant influence on Cr density only. Initial temperatures at current zero do not differ much for the same contact type (cf. Figure 9). The density slightly increases with longer arc duration. Stronger deviations in the density were found when the arc ignition position was varied. A higher electrode temperature leads consequently to higher density of chromium vapor in the case of the electrodes with position A. It amounts to about 1.1×10^{18} m^{-3} for starting position A immediately after current interruption, while the value for position B is about 9.5×10^{17} m^{-3}, even though the arc duration was longer. The results are similar for other time instants.

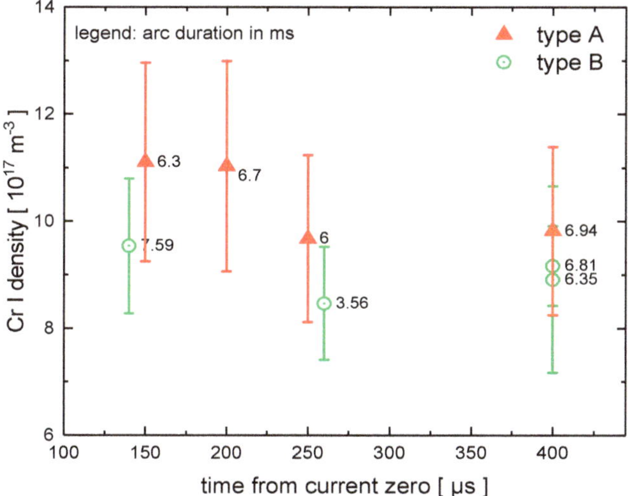

Figure 10. Cr I density measured after current zero crossing for different shots with contact systems of type A (red symbols) and B (green symbols). The numbers indicate the arc duration in ms.

4. Summary

The conducted study clarified that there is a clear influence of initiation behavior of the drawn arc on the electrode temperature, which is about 100 K (8%) higher in case of outer radial position measured directly after current interruption. Cr vapor density after current interruption is about 10% higher for outer radial positions. Therefore, it can be concluded that arc ignition in outer regions leads to higher local anode temperature and mean vapor density. Such behavior could potentially lead to increased electrode erosion, shorter lifetimes, and more restrikes (breakdown after current interruption). Future electrode design should take into account possible changes in electrode surface topology after significant current load.

Further investigations are planned to study the dependence of obtained results on material (electrode composition), as well other electrode design (such as AMF contacts).

Author Contributions: Conceptualization, S.G., A.L., E.D.T., F.G.; methodology, S.G., R.M., S.F.; software, S.F., S.G.; validation, E.D.T., F.G.; formal analysis, S.G.; investigation, S.G., D.G.; resources, D.G.; data curation, S.G.; writing—original draft preparation, S.G.; writing—review and editing, S.F., R.M., D.G., A.L., E.D.T. and F.G.; visualization, S.G., R.M. and S.F.; supervision, D.G. and F.G.; project administration, S.G. and A.L. All authors have read and agreed to the published version of the manuscript.

Funding: This research received no external funding.

Conflicts of Interest: The authors declare no conflict of interest.

References

1. Slade, P.G. *The Vacuum Interrupter: Theory, Design, and Application*; CRC Press: Boca Raton, FL, USA, 2008.
2. Giere, S.; Hellig, D.; Koletzko, M.; Kosse, S.; Rettenmaier, T.; Stiehler, C.; Wenzel, N. Vacuum interrupter unit for CO_2-neutral 170kV/50kA switchgear. In *ETG-Fb. 157: VDE-Hochspannungstechnik 2018*; 12.-14.11.2018; VDE Verlag GmbH: Berlin, Germany, 2018; pp. 28–31.
3. Lippmann, H.J. *Schalten im Vakuum, Physik und Technik der Vakuumschalter*; VDE Verlag GmbH: Berlin, Germany, 2003.
4. Schneider, H.N. Contact Structure for an Electric Circuit Interrupter. U.S. Patent 2,949,520, 19 August 1958.
5. Lake, A.A.; Reece, M.P. Improvements Relating to Vacuum Switch Contact Assemblies. UK Patent 997,384, 7 July 1965.

6. Hundstad, R.L. Contact Structures for Vacuum-Type Circuit Interrupters Having Cantilevered-Supported Annularly-Shaped Outer Arc-Running Contact Surfaces. U.S. Patent 3,845,262, 29 October 1974.
7. Lamara, T.; Hencken, K.; Gentsch, D. A Novel Vacuum Interrupter Contact Design for Improved High Current Interruption Performance Based on a Double-TMF Arc Control System. *IEEE Trans. Plasma Sci.* **2015**, *43*, 1798–1805. [CrossRef]
8. Dullni, E.; Schade, E. Investigation of high-current interruption of vacuum circuit breakers. *IEEE Trans. Electr. Insul.* **1993**, *28*, 607–620. [CrossRef]
9. Miller, H.C. Anode modes in vacuum arcs: Update. *IEEE Trans. Plasma Sci.* **2017**, *45*, 2366–2374. [CrossRef]
10. Schulman, M.B.; Schellekens, H. Visualization and characterization of high-current diffuse vacuum arcs on axial magnetic field contacts. *IEEE Trans. Plasma Sci.* **2000**, *28*, 443–451. [CrossRef]
11. Shang, W.; Dullni, E.; Fink, H.; Kleberg, I.; Schade, E.; Shmelev, D.L. Optical investigations of dynamic vacuum arc mode changes with different axial magnetic field contacts. *IEEE Trans. Plasma Sci.* **2003**, *31*, 923–928. [CrossRef]
12. Gentsch, D.; Shang, W. High-speed observations of arc modes and material erosion on RMF- and AMF-contact electrodes. *IEEE Trans. Plasma Sci.* **2005**, *33*, 1605–1610. [CrossRef]
13. Wolf, C.; Kurrat, M.; Lindmayer, M.; Wilkening, E.-D.; Gentsch, D. Optical investigations of high-current vacuum arc behavior on spiral-shaped and cup-shaped RMF-contacts. In Proceedings of the 55th IEEE Holm Conference on Electrical Contacts, Vancouver, BC, Canada, 14–16 September 2009; pp. 270–275.
14. Methling, R.; Franke, S.; Gortschakow, S.; Abplanalb, M.; Sütterlin, R.-P.; Delachaux, T.; Menzel, K.O. Anode surface temperature determination in high current vacuum arcs by different methods. *IEEE Trans. Plasma Sci.* **2017**, *45*, 2099–2107. [CrossRef]
15. Chaly, A.M.; Poluyanova, I.N.; Yakovlev, V.V.; Zabello, K.K.; Logatchev, A.A.; Shkol'nik, S.M. Experimental study of anode surface temperature after current zero for a range of current levels. In Proceedings of the 27th International Symposium on Discharges and Electrical Insulation in Vacuum (ISDEIV), Suzhou, China, 18–23 September 2016; pp. 1–4. [CrossRef]
16. Logachev, A.A.; Poluyanova, I.N.; Zabello, K.K.; Begal, D.I.; Shkol'nik, S.M. Analysis of cathode surface state and cathode temperature distribution after current zero of AMF-contacts. In Proceedings of the 28th International Symposium on Discharges and Electrical Insulation in Vacuum (ISDEIV), Greifswald, Germany, 23–28 September 2018; pp. 329–332.
17. Gortschakow, S.; Popov, S.; Khakpour, A.; Schneider, A.; Methling, R.; Franke, S.; Uhrlandt, D. Cu and Cr density determination during high-current discharge modes in vacuum arcs. In Proceedings of the 28th International Symposium on Discharges and Electrical Insulation in Vacuum(ISDEIV), Greifswald, Germany, 23–28 September 2018; pp. 181–184.
18. Horvath, B.; Lamara, T. Time-resolved optical resonant absorption spectroscopy of Cr metallic vapor in air using a broadband LED light source. *Plasma Sources Sci. Technol.* **2013**, *22*, 035006. [CrossRef]
19. Wang, H.; Wang, Z.; Liu, U.J.; Zhou, Z.; Wang, J.; Geng, Y.; Liu, Z. Optical absorption spectroscopy of metallic (Cr) vapor in a vacuum arc. *J. Phys. D Appl. Phys.* **2018**, *51*, 035203. [CrossRef]
20. Khakpour, A.; Popov, S.; Franke, S.; Kozakov, R.; Methling, R.; Uhrlandt, D.; Gortschakow, S. Determination of Cr density after current zero in a high-current vacuum arc considering anode plume. *IEEE Trans. Plasma Sci.* **2017**, *45*, 2108–2114. [CrossRef]
21. Gortschakow, S.; Khakpour, A.; Popov, S.; Franke, S.; Methling, R.; Uhrlandt, D. Determination of Cr density in a high-current vacuum arc considering anode activity. *Plasma Phys. Technol.* **2017**, *4*, 190–193. [CrossRef]
22. Methling, R.; Gorchakov, S.; Lisnyak, M.V.; Franke, S.; Khakpour, A.; Popov, S.; Batrakov, A.; Uhrlandt, D.; Weltmann, K.-D. Spectroscopic investigation of high-current vacuum arcs. In Proceedings of the 26th 2014 International Symposium on Discharges and Electrical Insulation in Vacuum (ISDEIV), Mumbai, India, 28 September–3 October 2014; pp. 221–224.

23. Günther, K.; Radtke, R. A proposed radiation standard for the visible and UV region. *J. Phys. E* **1975**, *8*, 371–376. [CrossRef]
24. Boxman, R.L.; Martin, P.J.; Sanders, D.M. *Handbook of Vacuum Arc Science and Technology*; Noyes: Hammonton, NJ, USA, 1995.

Publisher's Note: MDPI stays neutral with regard to jurisdictional claims in published maps and institutional affiliations.

Article

DC Current Interruption Based on Vacuum Arc Impacted by Ultra-Fast Transverse Magnetic Field

Ehsan Hashemi [1] and Kaveh Niayesh [2,*]

1 School of Electrical and Computer Engineering, Faculty of Engineering, University of Tehran, Tehran 1417466191, Iran; ehsanhashemi@ut.ac.ir

2 Department of Electric Power Engineering, Norwegian University of Science and Technology (NTNU), 7034 Trondheim, Norway

* Correspondence: kaveh.niayesh@ntnu.no

Received: 24 July 2020; Accepted: 4 September 2020; Published: 7 September 2020

Abstract: In this paper, the effect of an external ultrafast transverse magnetic field (UFTMF) on a vacuum arc in the diffused mode has been studied. According to the results of studies, a novel approach for making a zero-crossing in a DC arc current has been presented. Plasma voltage fluctuations of the vacuum arc, which are caused by UFTMF, have been investigated via finite element simulation and two-fluid description of plasma physics. By making an appropriate UFTMF through an external circuit, the arc current can be commuted successfully from the vacuum interrupter (VI) to a parallel capacitor and charge it up. In this way, a zero-crossing in the arc current can be achieved, and the current will be interrupted by the VI. Simulation results, which are supporting physical backgrounds for this analysis, have been presented in this paper while technological issues for industrial implementation of this concept have been discussed in detail.

Keywords: vacuum circuit breaker; vacuum arc; DC circuit breaker; current interruption; vacuum interrupter; magnetic field; plasma physics; zero-crossing

---

1. Introduction

Various studies on vacuum interrupters (VIs) have been performed in recent decades. VI has been considered as an excellent interrupting technology, which is able to interrupt high-frequency currents [1–3]. The effect of the magnetic field on the arc in VI has been investigated both experimentally and analytically for axial and transverse magnetic fields because of the vital role of the magnetic field in VI design. This research gives an approach to VI designers to improve the structure of VI contacts and materials. The interpretation of the plasma behavior under magnetic fields is limited to the selected physical model. Earlier investigations [4,5] applied a constant temperature model. Other researchers [6] used a magneto-hydro-dynamic model, which is a single fluid model. In [7,8], researchers considered the movement of a constricted arc column, which is imposed by Lorentz force and move freely according to fluid dynamic equations.

Extensive experimental investigations were performed in [9,10], where the effect of the magnetic field on the cathode spot displacement was reported. Interesting experimental observations were reported in [11] in which an increment in arc voltage by using an external magnetic field was studied.

In the case of DC current breaking, there are several technologies, which are currently used. The passive inductive-capacitive oscillator circuit and active current injection are the applicable technologies now. As described in [12,13], there is a basic trend to find and present new technologies to make an artificial current zero-crossing and optimize the structure of existing DC circuit breakers in terms of technical complexity, weight, and cost.

This paper introduces a novel idea for making a zero-crossing in DC arc currents by using an ultrafast transverse magnetic field (UFTMF). This idea is based on the transient disturbances in plasma caused by imposed external UFTMF to a low-pressure plasma chamber such as a VI during arcing. The main disturbance is an immediate response of electrons to the change of plasma medium characteristics and lagged movement of heavy charged particles.

Simulation of the phenomenon needs a two-fluid plasma model, which ensures different fluxes and temperatures of electrons and ions. By this analysis, expected disturbances in copper vapor plasma have been investigated and the results have been presented. In this paper, a similar plasma model has been employed to simulate the copper vapor plasma, which is the medium of commercial VIs, and a parallel capacitor with specific characteristic considered as an external circuit of VI. It is observed that a fast increment in arc voltage caused by external UFTMF leads to a successful commutation of arc current from VI to the parallel capacitor. This resulted in a zero-crossing in the arc current, which can be considered as a successful current interruption in VI because of the high-frequency interruption capability of VI technology.

In the next sections, the physical model and simulation results for copper vapor plasma in VI have been presented, and the idea that shows how an ultrafast transverse magnetic field can make a zero-crossing in the arc current has been discussed.

2. Physical Model and Geometry Simulation

Two disks with a radius of 5.5 mm were considered as contacts of the interrupter. The maximum contact separation distance was assumed to be 12 mm as what is implemented in commercial VIs. A DC voltage source with an amplitude of 1000 V supplied the current of 50 A through a 20 Ohm resistance, which is placed in series with the interrupter.

The employed physical model was as described in [14]. The continuity equation of electrons, mass conservation equations of heavy species were considered as well as the heat transfer and Maxwell equations [15–19]. These equations are as follows:

$$\frac{\partial n_{electron}}{\partial t} + \nabla\cdot\vec{\Gamma}_{electron} = RR_{electron} - (\vec{u}\cdot\nabla)\, n_{electron}$$
$$\vec{\Gamma}_{electron} = -(\vec{\mu}_{electron}\cdot\vec{E})n_{electron} - \vec{D}_{electron}\cdot\nabla n_{electron} \tag{1}$$

$$\frac{\partial n_{e,energy}}{\partial t} + \nabla\cdot\vec{\Gamma}_{e,energy} + \vec{E}\cdot\vec{\Gamma}_{e,energy} = RR_{e,energy} - (\vec{u}\cdot\nabla)\, n_{e,energy}$$
$$\vec{\Gamma}_{e,energy} = -(\vec{\mu}_{e,energy}\cdot\vec{E})n_{e,energy} - \vec{D}_{e,energy}\cdot\nabla n_{e,energy},\ \vec{\mu}_{e,energy} = \tfrac{5}{3}\vec{\mu}_{electron} \tag{2}$$
$$\vec{D}_{e,energy} = \vec{D}_{electron}\, T_{electron}$$

$$\rho_{hs}\frac{\partial M_f}{\partial t} = RR_a + \nabla\cdot(\rho_{hs}M_f D_a^m \vec{\nabla}\ln(M_f) + D_a^T\vec{\nabla}\ln(T) + \rho_{hs}M_f z_{ion}u_{ion}\vec{E}) \tag{3}$$

$$\rho_{hs}C_P\frac{\partial T}{\partial t} + \rho_{hs}C_P\,\vec{u}\cdot\vec{\nabla}T = \nabla\cdot(k\vec{\nabla}T) + Q_s \tag{4}$$

$$\nabla\cdot\vec{E} = \frac{e(zn_{ion} - n_{electron})}{\varepsilon_0} \tag{5}$$

$$\vec{j}_{total} = e\,(zn_{ion}\vec{v}_{ion} - n_{electron}\vec{v}_{electron}) \tag{6}$$

The continuity equation of electrons was described as Equation (1). In this equation $n_{electron}$, u, $D_{electron}$, $RR_{electron}$, $\mu_{electron}$, and E represent the electron number density, convection velocity of fluid, diffusion coefficient and reaction rate of electron production, electron mobility, and electric field, respectively.

Energy conservation of electrons was described as Equation (2). In this equation $n_{e,energy}$, $D_{e,energy}$, and $\mu_{e,energy}$ represent the electron energy density, electron energy diffusion coefficient, and electron energy mobility, respectively. The rate of change in electron energy according to inelastic impacts between electrons and heavy species was described by $RR_{e,energy}$.

Mass conservation of the heavy particles was described in Equation (3). In this equation M_f, ρ_{hs}, RR_{hs}, ${D_{hs}}^m$, ${D_{hs}}^T$, z_{ion}, u_{ion}, and T are the mass fraction, fluid density, production rate of heavy species, drift diffusion coefficient, thermal diffusion coefficient, ion charge number, ion mobility, and temperature of the heavy species, respectively. The type of heavy particle—neutral atom, excited atom, or ion—was presented by index 'a'.

Heat transfer in a fluid medium was described in Equation (4). In this equation C_p, k, and Q_s are the heat capacity, thermal conductance, and heat source, respectively.

Maxwell description for the electric field was presented in Equation (5). In this equation, n_{ion} and ε_0 are the ion number density and vacuum permittivity respectively.

Equation (6) represents the total current density flowing through the arc in which n_{ion}, e, $v_{electron}$, and v_{ion} are the ion density, electric charge of electron, electron velocity, and ion velocity, respectively.

The considered particles in copper vapor plasma were electron, Cu, Cu^*, $Cu+$, and $Cu++$, which are major species in this type of plasma [6]. The reaction rates between electrons and heavy species, which cause ionization and generation of the ions and excited particles, were calculated by using the impact cross-section data. In this way, an effective cross-section of electrons was calculated as a function of the electron energy distribution function. Required data were extracted from [15,20].

The two main parameters for calculation of reaction rates were the collision cross section value of electrons and the Townsend ionization coefficient versus electron energy. The assumption of the Maxwellian distribution function for electron energy was taken into account [15,16]. The considered particles in this analysis and chemical reactions between electrons, ions, and atoms were expressed as Equations (7)–(12).

$$e + Cu \Rightarrow e + Cu \ (Elastic\, reaction) \tag{7}$$

$$e + Cu \Rightarrow e + Cu^* \ (Excitation) \tag{8}$$

$$e + Cu \Rightarrow 2e + Cu^+ \ (Ionization) \tag{9}$$

$$e + Cu^+ \Rightarrow 2e + Cu^{++} \ (Ionization) \tag{10}$$

$$e + Cu^* \Rightarrow 2e + Cu^+ \ (Stepwise\, Ionization) \tag{11}$$

$$Cu + Cu^* \Rightarrow e + Cu^+ + Cu \ (Penning\, Ionization) \tag{12}$$

2.1. Boundary Condition of the Cathode

According to [6], 70 percent of the cathode surface was considered as an active area for electric current injection. A diffused mode arc was taken into account when a 50 A plasma channel existed between the contact area and the magnetic field of the plasma current was unable to constrict the arc column. In this situation, variations of plasma parameters in an arc medium were considered while the arc–cathode interaction area was assumed to be unchanged.

Two other assumptions to make up the cathode boundary in the equations were secondary electron emission and considering the cathode to be hot enough to emit the electrons with respect to the electric field of the sheath region.

Equations describing the behavior of the cathode in an electron flux issue are as below:

$$\begin{aligned} &\vec{n} \cdot \vec{\Gamma}_{electron} = \tfrac{1}{2} v_{electron,th}\, n_{electron} + n_{electron} \left(\vec{\mu}_{electron} \cdot \vec{E}\right) \cdot \vec{n} - \\ &\left(\gamma_{ion}\left(\vec{n} \cdot \vec{\Gamma}_{ion}\right) + \vec{n} \cdot \vec{\Gamma}_{electron,injected}\right) \end{aligned} \tag{13}$$

$$\vec{n} \cdot \vec{\Gamma}_{e,energy} = \tfrac{1}{2} v_{electron,th}\, n_{e,energy} + n_{e,energy}\,(\vec{\mu}_{e,energy} \cdot \vec{E}) \cdot \vec{n} - \\ (\gamma_{ion}\, \varepsilon_{ion}\,(\vec{n} \cdot \vec{\Gamma}_{ion}) + \varepsilon\, \vec{n} \cdot \vec{\Gamma}_{electron,injected}) \tag{14}$$

In Equations (13) and (14), $v_{electron,th}$, γ_{ion}, ε_{ion}, ε, $\Gamma_{electron,injected}$, and n are the electron thermal velocity, secondary electron emission coefficient, mean energy of the ion secondary emitted electrons, mean energy of thermally emitted electrons, total injected electron flux, and normal vector, respectively. The total injected electron flux was determined by the current flowing through the plasma by the external circuit as described later.

The cathode acted on ions and excited atoms as Equations (15)–(17). According to these equations, the cathode provides electrons to neutralize the positive ions and absorbs the excitation energy of the excited atom

$$Cu^{+} \Rightarrow Cu \tag{15}$$

$$Cu^{++} \Rightarrow Cu \tag{16}$$

$$Cu^{*} \Rightarrow Cu \tag{17}$$

In the heat transfer phenomenon, the cathode surface absorbs the heat flux from plasma and the temperature of the cathode material rises. Indeed, the cathode acts as the outgoing path of the heat flux.

2.2. Boundary Condition of the Anode

The anode acts as a perfect sink for electrons [6]. In comparison with the cathode, the perfect sink is a boundary condition in which the total electron flux on the anode is equal to the sum of the drift and thermal electron fluxes with no sources. In this research, the temperature of the anode will not increase up to the melting point thus the anode will remain inactive during the process and no heavy species are emitted from the anode. Under these conditions, the anode acts as an electron wall, which provides electrons for positive ions hitting the anode, and converts them to neutral atoms. Equations (15)–(17) are valid for the anode too.

The performance of the anode in heat transfer phenomena is the same as the cathode.

2.3. Boundary Conditions of Chamber Walls

The walls of the plasma chamber act for heavy species based on Equations (15)–(17). The effect of the walls on electrons is the same as the anode with the difference that in electrical equations the wall acts as a surface charge accumulator. These effects are described as:

$$\frac{\partial \rho_s}{\partial t} = \vec{n} \cdot \vec{j}_{electron} + \vec{n} \cdot \vec{j}_{ion} \\ \vec{n} \cdot \vec{D} = -\rho_s \tag{18}$$

In this equation, ρ_s, $j_{electron}$, j_{ion}, D, and n are the surface charge density, electron current density, ion current density, electrical displacement, and normal vector, respectively.

2.4. Interaction between the External Transverse Magnetic Field and Low-Pressure Plasma

In the present research, the effect of the magnetic field on the electron movement was considered as a change in the mobility tensor of electrons. The tensor expression of electron mobility was replaced by a scalar one to model the magnetized plasma parameters. Equation (19) represents the mobility tensor.

$$\mu_{magnetized}^{-1} = \begin{bmatrix} \mu^{-1} & -B_z & B_y \\ B_z & \mu^{-1} & -B_x \\ -B_y & B_x & \mu^{-1} \end{bmatrix} = \begin{bmatrix} \mu_{xx} & \mu_{xy} & \mu_{xz} \\ \mu_{yx} & \mu_{yy} & \mu_{yz} \\ \mu_{zx} & \mu_{zy} & \mu_{zz} \end{bmatrix} \tag{19}$$

In Equation (19) $\mu_{magnetized}$, μ, B_x, B_y, B_z, μ_{xx}, μ_{yy}, μ_{zz}, μ_{xy}, μ_{xz}, and μ_{yz} are the mobility tensor, mobility of electron in non-magnetized plasma, magnetic flux density along the x, y, and z axes, and elements of the mobility tensor, respectively. In the present research, an external transverse magnetic field was applied to the current path perpendicularly. In 2D modeling, if the plasma was assumed to be placed in the X–Y plane, a desirable external magnetic field should be along the Z-axis. This magnetic field pushes the plasma to one side of the chamber depending on the magnetic field direction. Cylindrical coordinates and axial symmetry were used in this paper to reach a convergent solution. By this assumption, moving of plasma to one side of the chamber changes to concentration or dispersion of the plasma.

The plasma chamber was assumed to carry the current of 50 A in the diffused mode. Copper vapor plasma with the pressure of 10 mbar was assumed between contacts [14].

The space between the contacts was exposed to a transverse magnetic field. First, the simulation results in the case of a small magnetic field (e.g., 20 mT) were presented. Then, the impact of stronger magnetic fields (e.g., 1 T) on the arc voltage behavior was discussed. Finally, the contribution of the arc voltage fluctuation to a successful current commutation was investigated.

3. Simulation Results and the Proposed System

A finite element (FEM) simulation was performed to study the effect of UFTMF on the arc voltage and other plasma characteristics. The described geometry of contacts and plasma medium was implemented for calculations in a 2D geometry with the assumption of axial symmetry. This symmetry completely expressed the plasma characteristics in the absence of a transverse magnetic field. There may be doubt about how a transverse magnetic field in axial symmetry, φ-direction, will be physically realized. The axial symmetry was employed just to reach a convergent numerical simulation for ultrafast transients in plasma parameters. The transformation from cylindrical (r, φ, and z) coordinates to Cartesian (x, y, and z) coordinates is straightforward. A magnetic field in the φ-direction acts on the plasma in the RZ plane in the same way as a magnetic field in the z-direction does in the XY plane.

In this simulation, a homogeneous background distribution of electrons, copper ions, and neutral copper atoms was assumed as the initial condition of the plasma. By considering boundary conditions for the anode and the cathode, the voltage was applied to the arc chamber and after 1 ms plasma reached the steady state with an electric current of 50 A. Then, an external UFTMF was applied to the arc chamber with a value of 20 mT and a rise-time of 500 nSec. The disturbances in electron distribution and the energy level of electrons and the electron temperature in plasma have been illustrated in Figure 1 at different time steps. As depicted in Figure 1, electrons were trapped behind the magnetized region because of an abrupt change in the mobility tensor of electrons. The peak value of electron density changed from 8.59×10^{19} to 1.25×10^{20} m^{-3} within 500 μSec.

A net space charge appeared near the magnetized region, and so there would be a strong electric field compared to what existed in the neutral plasma medium. This electric field caused electrons to gain more energy. That is why there were high-temperature electrons as depicted in Figure 1e–h. Indeed, there was an increment in the electron temperature from 1.38 to 2.29 eV. These high-temperature electrons caused high reaction rates and high power dissipation in the plasma medium. Figure 1i–l shows the variations of power dissipation in plasma.

Electron flux and related streamlines have been illustrated in Figure 2. The trajectory of electrons was broken when electrons entered the magnetized region. If the magnetic field region contains contacts, what may happen in reality is a high electric field region that forms from the cathode to the anode. This instantaneous disturbance in the electron flux and consequently electron density and net space charge was responsible for the instantaneous extreme value of the electric field in the plasma medium and the increment in arc voltage.

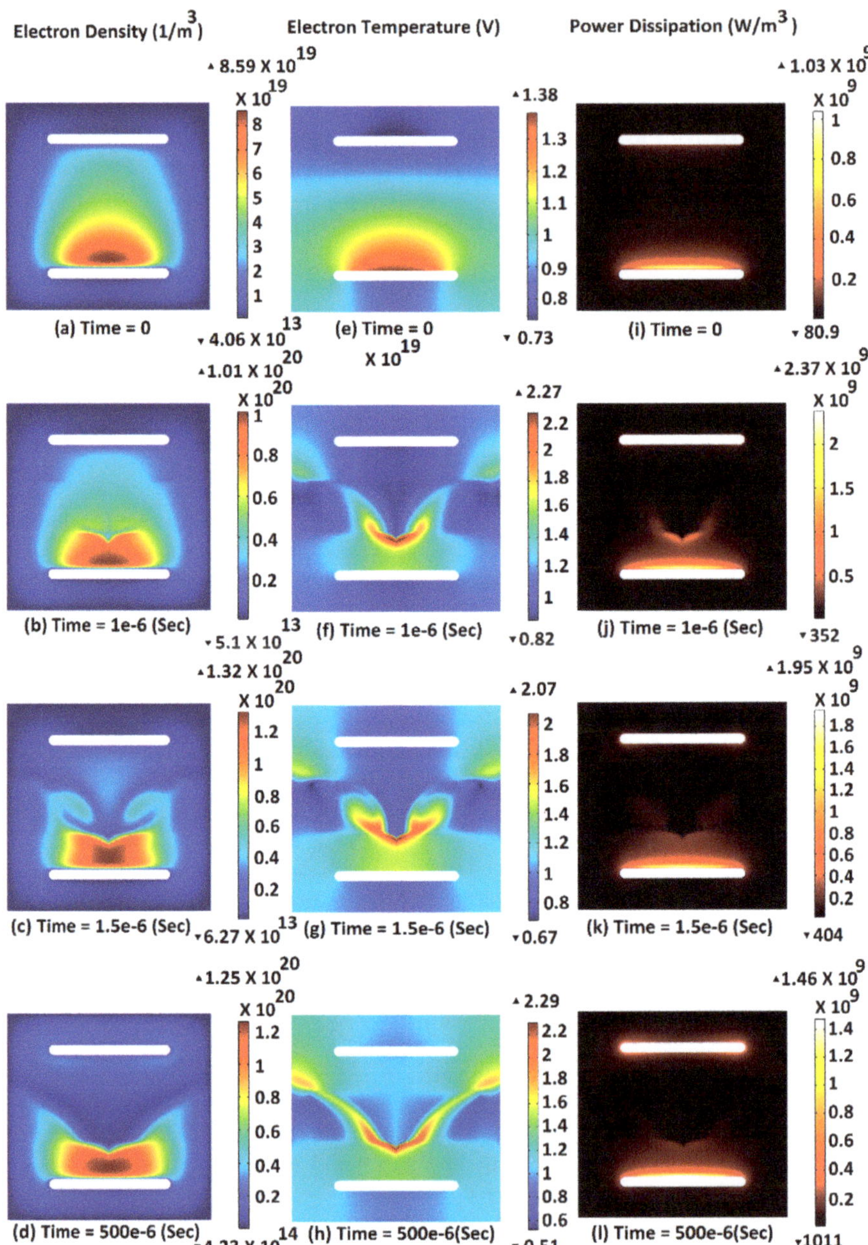

Figure 1. Distribution of electron, electron temperature, and power dissipation in plasma at different time steps: (**a**,**e**,**i**) initial moment, (**b**,**f**,**j**) 1 µSec, (**c**,**g**,**k**) 1.5 µSec and (**d**,**h**,**l**) 500 µSec after application of 20 mT ultrafast transverse magnetic field (UFTMF).

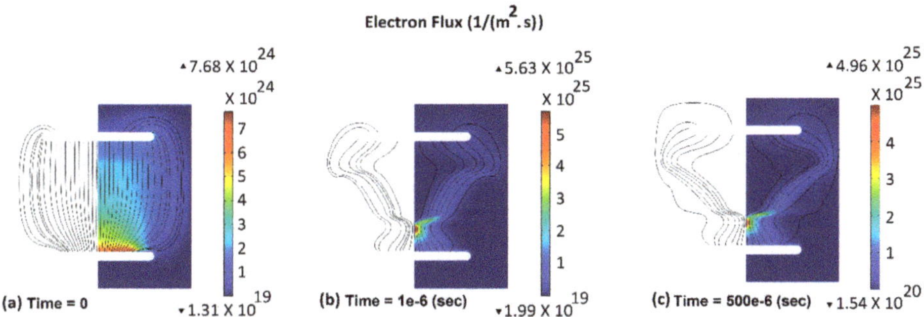

Figure 2. Distribution of the electron flux at different time steps: (**a**) initial moment, (**b**) 1 μSec and (**c**) 500 μSec after application of 20 mT UFTMF.

Figure 3a shows the electric potential distribution across the central axis of the plasma chamber at different time steps 0, 1 μSec, 1.5 μSec, 30 μSec, and 500 μSec after application of UFTMF. As it can be seen in Figure 3a, there was a dramatic increment in arc voltage in a short time duration. Actually, arc voltage jumped from 5.9 to 10.1 V within 1 μSec. This instantaneous increment in arc voltage caused a transient voltage change with a rate of rise of 4.2 MV/sec. In this situation, a strong electric field region appeared near the magnetized region. The difference between the density of the electrons and ions, which was caused by the applied UFTMF, was responsible for this extreme electric field in the plasma medium. This region was about 3 mm from the cathode where there was an approximately zero electric field in the plasma medium before application of UFTMF. This strong electric field was the main reason for high-temperature electrons, high reaction rates, and the high power dissipation mentioned before.

In Figure 3b, potential distribution along the arc axis in the diffuse mode arc with the current of 15 kA has been depicted [6], which is rarely mentioned in the literature. As illustrated in Figure 3b from the perspective of the potential distribution, the arc consists of four major regions: cathode spots, cathode jet mixing, the plasma bulk, and the anode sheath [6]. The considered model for the diffuse mode vacuum arc in the present paper had mathematically modeled only three regions: cathode jet mixing, plasma bulk, and anode sheath because the focus was on what happened in plasma due to UFTMF and mechanisms of forming and extinguishing of cathode spots was not considered. Another issue in this paper is the level of the arc current. A voltage drop along the plasma bulk region was proportional to the arc current. Thus, a voltage drop of 8 V for the arc current of 15 kA in Figure 3b will disappear in Figure 3a. By this description and comparing Figure 3a,b, it can be seen that if the voltage drop across the cathode spot region and plasma bulk was neglected from Figure 3b, there would be a trend similar to what has been presented in Figure 3a at "Time = 0" and this similarity validates the physical model, which was used in this paper.

Observation of such a high rate of rise in arc voltage, 4.2 MV/sec, illustrates the idea that a higher value UFTMF will give a sufficient rate of rise in arc voltage to commutate the arc current to a parallel capacitor. Another simulation with 1 T UFTMF was performed to investigate this idea, and the results were as expected.

At the first step, no parallel capacitor was assumed in the circuit, and an UFTMF of 1 T was applied to the arc. An increment of 20 times in the arc voltage appeared, as shown in Figure 4.

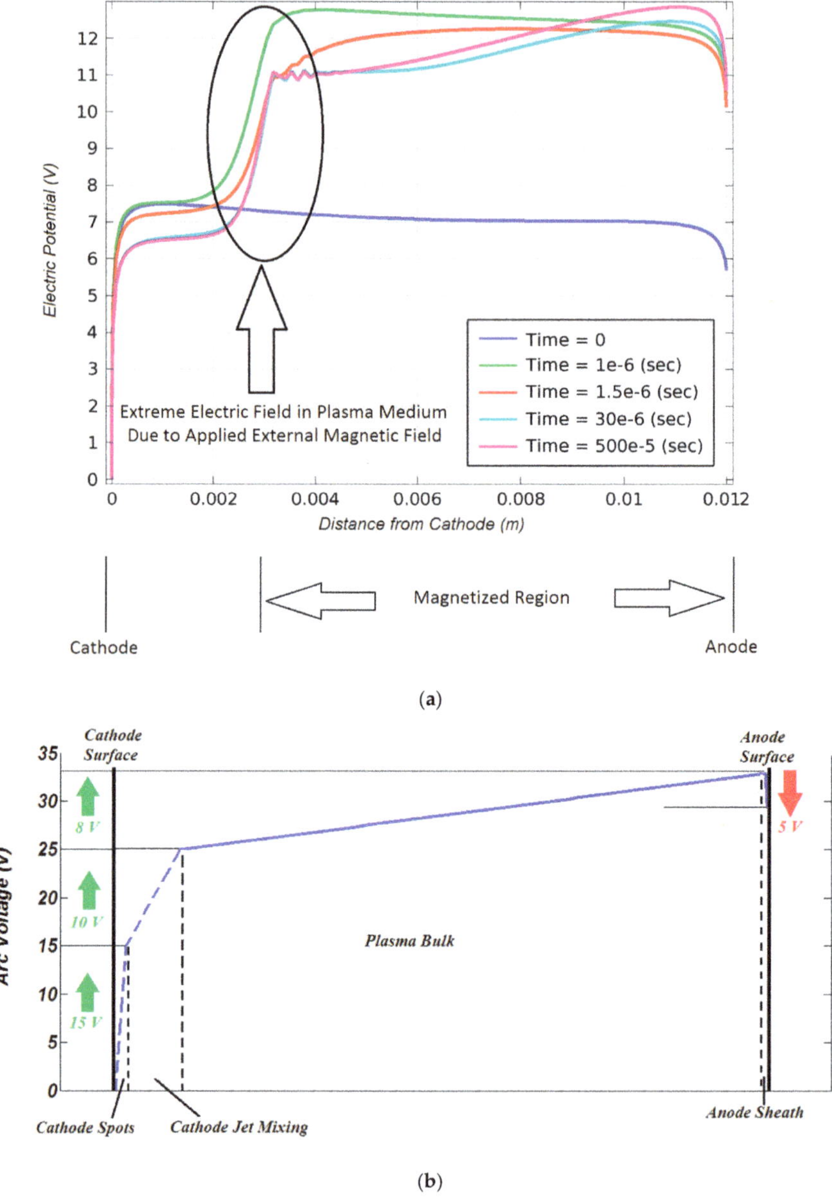

(a)

(b)

Figure 3. (**a**) Distribution of the electric potential across the symmetry axis of the plasma chamber at time steps: 0 μSec, 1 μSec, 1.5 μSec, 30 μSec, and 500 μSec after application of 20 mT UFTMF. (**b**) Distribution of the electric potential from the cathode to the anode according to the 15 kA arc imposed by the axial magnetic field [6].

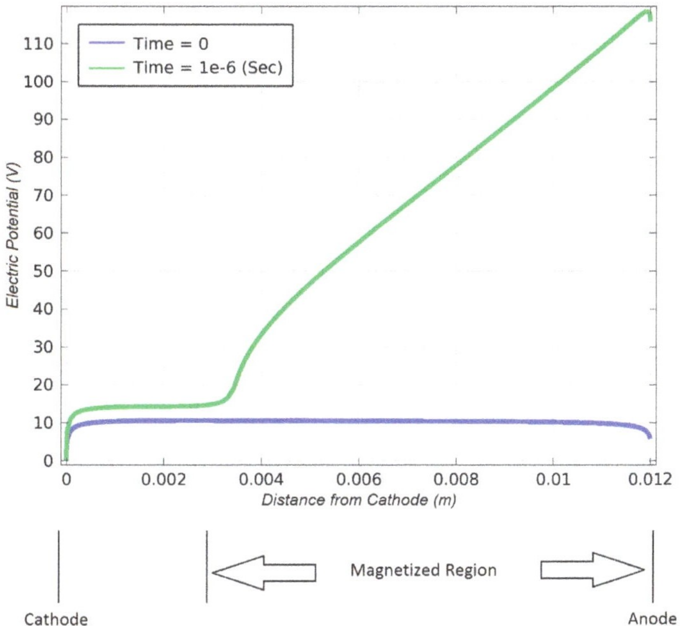

Figure 4. Distribution of the electric potential across the symmetry axis of the plasma chamber at time steps: 0 and 1 µSec after application of 1 T UFTMF.

Figure 4 shows that the arc voltage increased from 5.9 to 117 V by the application of an UFTMF of 1 T within just 1 µSec. A value of the parallel capacitor as illustrated in Figure 5 was calculated by performing numerous simulations with a focus on the commutation of the arc current. Various values of the parallel capacitor were chosen and simulated simultaneously with a plasma analysis in a range of a few pF to several µF. An appropriate value for successful commutation was calculated as 200 nF.

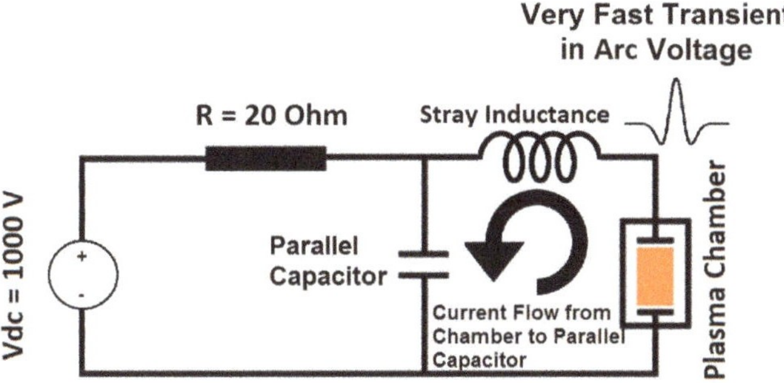

Figure 5. The external circuit of the plasma chamber, which is equipped by a parallel capacitor.

As it is mentioned before, arc current commutation from the plasma chamber to the parallel capacitor will be considered as an interruption of the arc current if the interrupter is capable of a high-frequency current interruption. Waveforms of currents flowing through the arc and the parallel capacitor were calculated and depicted in Figure 6.

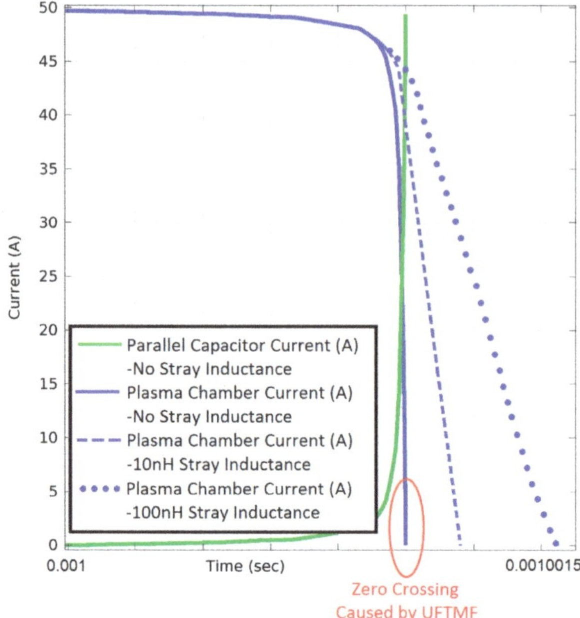

Figure 6. Waveforms of the plasma chamber and parallel capacitor current after application of an UFTMF of 1 T. Successful commutation is achieved with the 200 nF parallel capacitor.

As it can be seen in Figure 6, a successful current commutation was achieved by placement of a 200 nF capacitor in parallel to the plasma chamber, where a zero-crossing in the arc current was observed. In the commutation process, the impedance of the parallel capacitor was much lower than the source side because of the high-frequency content of the arc voltage. The capacitor will charge up to a higher voltage than the initial arc voltage because of the direction of the electric current. By the charging process, the arc voltage will be restrained by the capacitor and a high arc voltage of 100 V as expressed in Figure 3 will not appear. Actually, the voltage of the capacitor scales up from 5.9 to 13.7 V. In addition, the effect of a stray inductance of the main circuit between the interrupter and parallel capacitor was considered for three values of 0, 10, and 100 nH. As depicted in Figure 6, by increment in stray inductance, a zero crossing time will occur with more delay and a zero crossing may be lost with large values of stray inductance. A design with a minimum value of stray inductance in this commutation system will be one of the key points.

The mentioned performance of this circuit under strong UFTMF expresses that successful interruption of the DC current will be accessible by this method while the transient recovery voltage across the interrupter will be limited by the parallel capacitor. There are, however, some technological issues, which have been discussed in the following.

4. Technological Discussions

Provision of UFTMF especially in medium-voltage power networks is the first challenge of this system. The main source of such high magnitude and an ultrafast magnetic field is a coil with the perpendicular axis to the current path. In addition, the coil should present a low value of inductance. In this system, there is a need to generate an ultrafast rising magnetic field, and consequently, there is a need to the ultrafast rising current in the coil. If the inductance of the coil is too high, the rise time of the current will increase, and UFTMF will not be accessible. The energy source to feed the coil may be a capacitor, which is charged to an appropriate voltage, and it can provide a sufficient amount of energy to increase the current in the coil. Capacitance and inductance of this system will satisfy the rise time of the electric current in the coil.

As illustrated in Figure 7, a two-turn coil was placed beside a VI. This coil was considered with two individual single turn coils on both sides of the VI, and the connecting path between two coils consisted of two symmetric parallel paths. These two parallel paths ensured the transverse magnetic field with the minimum axial component.

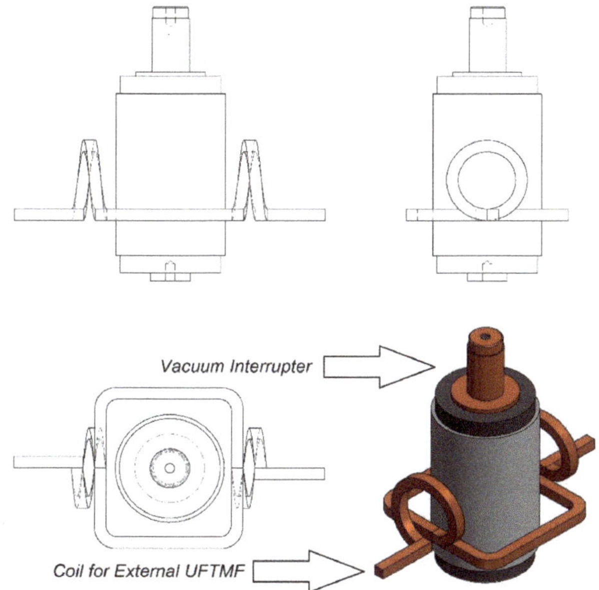

Figure 7. A vacuum interrupter (VI) equipped with an external UFTMF coil.

An electric current of 200 kA with a rise-time of 1 μSec flowing through the coil ensured UFTMF with a rate of rise of 10^6 T/sec.

Distribution of the magnetic flux density and related arrow-lines inside the arcing medium of VI has been depicted in Figure 8 for the mentioned coil current and by the use of a finite element simulation.

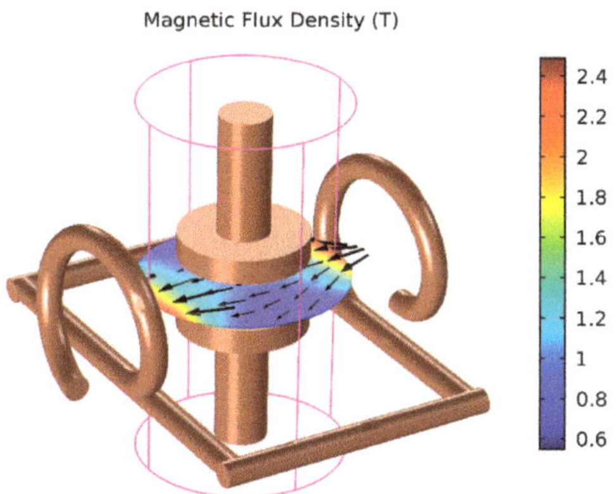

Figure 8. Distribution of the magnetic flux density amplitude generated by an electric current of 200 kA with 1 μSec rise-time.

A charged capacitor, which is capable of providing pulsed current to the coil, will be an appropriate choice of energy source for this system. The capacitance and charging voltage of the capacitor can be determined by knowing the inductance of the coil. Finite element simulation of the coil and calculation of the magnetic field and magnetic energy distribution expressed that the inductance of the coil was 0.2 μH. The capacitance and charging voltage of the capacitor were calculated by considering an inductance value of 0.2 μH, a peak current of 200 kA, and a rise time of 1 μSec. The calculated capacitance was 2 μF and the charging voltage was 60 kV. These values satisfied the necessary electric current for the coil to reach the mentioned criteria. A schematic diagram of the feeding circuit is depicted in Figure 9.

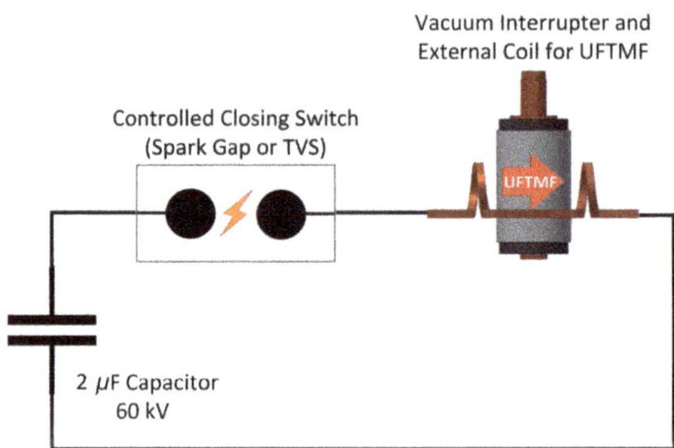

Figure 9. Schematic diagram of an electric current feeding circuit.

As illustrated in Figure 9, a closing switch capable of passing 200 kA is required. Possible candidates could be controlled spark gap closing switches or triggered vacuum switches (TVS). The breakdown voltage of the gap can be controlled by gap distance and gas pressure of the spark gap [21,22].

One of the big challenges in this technology is penetrating the magnetic field into the arc chamber. Electric contacts of VI and the switching gap are surrounded by the vapor shields [23]. According to the electromagnetic theory, it is expected that UFTMF will be damped by the metallic vapor shields because of its high-frequency content and as a consequence of large induced currents inside the shield. Conventional vapor shields act as a perfect electromagnetic shield for such UFTMF and so magnetic field damps dramatically inside the shield. This problem should be addressed by changing the structure of the vapor shield. One possible approach to resolve this problem is the replacement of the arc shield with a similar one with a different material. The use of one-piece ceramic vacuum interrupters as suggested in [24] could be considered in this context. The tolerable amount of a metal coating on an isolating material (e.g., due to the deposition of metal vapor on the ceramic shield) is dependent on the frequency range of UFTMF. Since the highest frequency content of UFTMF is 500 kHz, a coating layer of 10–20 μm will be completely transparent to the UFTMF. In addition, the challenge may be overcome by making specific cut-lines on the vapor shield and using an overlapping structure.

5. Conclusions

The system proposed in this paper can be an innovative concept to introduce new technology for HVDC circuit breakers and fault current limiters. Imposing an UFTMF with the magnitude of 1 T and the rise time of 1 μSec or less to a modified commercial VI led to a considerable disturbance in charged particles' spatial distribution and consequently to a transient increment in arc voltage. The electric current of VI will be commuted successfully to an appropriate parallel capacitive external circuit, and a zero-crossing in the electric current of VI will be achieved. All of these results were calculated by the finite element method based simulation of copper vapor plasma in VI while the interaction of plasma and UFTMF was considered.

This concept of current interruption introduces an alternative method for the interruption of electric currents when there is no zero-crossing, or there is a long delay time in reaching the zero-crossing. DC current in the HVDC power network and AC current of generators in the HVAC power etwork are the main cases of these applications. Another application of the proposed system is in the HVAC power network where there is a need to have an immediate current interruption. Fault current limiters are the main examples of this situation.

The main technological issues—provision of UFTMF, feeding circuit, and penetration of UFTMF into contact area of VI—were discussed. This novel method will definitely encounter other technological limitations such as necessary variations in the structure of VI, contact surface treatment for the better high-frequency current interruption, UFTMF positioning, etc. Thus, the presented results of this paper call for future experimental validations.

Author Contributions: Conceptualization, E.H. and K.N.; methodology, E.H. and K.N.; software, E.H.; validation, E.H.; formal analysis, E.H.; investigation, E.H. and K.N.; resources, E.H.; writing—original draft preparation, E.H.; writing—review and editing, E.H. and K.N.; visualization, E.H.; supervision, K.N.; funding acquisition, K.N. All authors have read and agreed to the published version of the manuscript.

Funding: This research received no external funding. The Article Processing Charges (APC) was funded by NTNU.

Conflicts of Interest: The authors declare no conflict of interest.

References

1. Slade, P.G. *The Vacuum Interrupter, Theory, Design, and Application*; CRC Press: Boca Raton, FL, USA, 2008.
2. Garzon, R.D. *High Voltage Circuit Breakers: Design and Applications*, 2nd ed.; Marcel Dekker: New York, NY, USA, 2002.
3. Niayesh, K.; Runde, M. *Power Switching Components, Theory, Applications and Future Trends*; Springer: Berlin/Heidelberg, Germany, 2017.
4. Boxman, R.L.; Goldsmith, S.; Izraeli, I.; Shalev, S. A Model of multicathode-spot vacuum arc. *IEEE Trans. Plasma Sci.* **1983**, *11*, 138–145. [CrossRef]
5. Keidar, M.; Beilis, I.; Boxman, R.; Goldsmith, S. Potential and current distribution in the interelectrode gap of the vacuum arc in a magnetic field. In Proceedings of the XVIIth International Symposium on Discharges and Electrical Insulation in Vacuum, Berkeley, CA, USA, 21–26 July 1996; pp. 146–150.
6. Schade, E.; Shmelev, D. Numerical simulation of high-current vacuum arcs with an external axial magnetic field. *IEEE Trans. Plasma Sci.* **2003**, *31*, 890–901. [CrossRef]
7. Shmelev, D.; Delachaux, T. Physical modeling and numerical simulation of constricted high-current vacuum arcs under the influence of a transverse magnetic field. *IEEE Tran. Plasma Sci.* **2009**, *37*, 1379–1385. [CrossRef]
8. Dullni, E.; Schade, E.; Shang, W. Vacuum arcs driven by cross-magnetic fields (RMF). *IEEE Trans. Plasma Sci.* **2003**, *31*, 902–908. [CrossRef]
9. Chaly, A.; Logatchev, A.; Zabello, K.; Shkol'nik, S. High-current vacuum arc appearance in nonhomogeneous axial magnetic field. *IEEE Trans. Plasma Sci.* **2003**, *31*, 884–889. [CrossRef]
10. Zabello, K.; Barinov, Y.; Chaly, A.; Logatchev, A.; Shkol'nik, S. Experimental study of cathode spot motion and burning voltage of low-current vacuum arc in magnetic field. *IEEE Trans. Plasma Sci.* **2005**, *33*, 1553–1559. [CrossRef]
11. Emtage, P.; Kimblin, C.; Gorman, J.; Holmes, F.; Heberlein, J.; Voshall, R.; Slade, P. Interaction between vacuum arcs and transverse magnetic fields with application to current limitation. *IEEE Trans. Plasma Sci.* **1980**, *8*, 314–319. [CrossRef]
12. Franck, C.M. HVDC circuit breakers: A review identifying future research needs. *IEEE Trans. Power Deliv.* **2011**, *26*, 998–1006. [CrossRef]
13. Jadidian, J. A compact design for high voltage direct current circuit breaker. *IEEE Trans. Plasma Sci.* **2009**, *37*, 1084–1091. [CrossRef]
14. Hashemi, E.; Niayesh, K.; Mohseni, H. Effect of transverse magnetic field on low-pressure argon discharge. *Turk. J. Electr. Eng. Comput. Sci.* **2016**, *24*, 4957–4969. [CrossRef]

15. Hagelaar, G.J.M.; Pitchford, L.C. Solving the Boltzmann equation to obtain electron transport coefficients and rate coefficients for fluid models. *Plasma Sources Sci. Technol.* **2005**, *14*, 722–733. [CrossRef]
16. Bellan, P. *Fundamentals of Plasma Physics*; Cambridge University Press: Cambridge, UK, 2008.
17. Bird, R.; Stewart, W.; Lightfoot, E. *Transport Phenomena*, 2nd ed.; Wiley: New York, NY, USA, 2002.
18. Kee, R.; Coltrin, M.; Glarborg, P. *Chemically Reacting Flow Theory and Practice*; Wiley: New York, NY, USA, 2003.
19. Neufeld, P.; Janzen, A.; Aziz, R. Empirical equations to calculate 16 of the transport collision integrals for the Lennard-Jones potential. *J. Chem. Phys.* **1972**, *57*, 1100–1102. [CrossRef]
20. Tkachev, A.N.; Fedenev, A.A.; Yakovlenko, S.T. Townsend coefficient, escape curve and fraction of runaway electrons in copper vapor. *Laser Phys.* **2007**, *17*, 775–781. [CrossRef]
21. Williams, P.F.; Peterkin, F.E. *Gas Discharge Closing Switches*; Plenum Press: New York, NY, USA, 1990.
22. Niayesh, K.; Hashemi, E.; Agheb, E.; Jadidian, J. Subnanosecond breakdown mechanism of low-pressure gaseous spark gaps. *IEEE Trans. Plasma Sci.* **2008**, *36*, 930–931. [CrossRef]
23. *VD4 Medium Voltage Vacuum Circuit Breakers Technical Catalog*; Rev. X; ABB Inc.: Zurich, Switzerland, 2018.
24. Falkingham, L.T. The design and development of the shieldless vacuum interrupter concept. In Proceedings of the 22nd International Symposium on Discharges and Electrical Insulation in Vacuum, Yalta, Ukraine, 27 September–1 October 2004; pp. 430–433.

Article

Research on Vacuum Arc Commutation Characteristics of a Natural-Commutate Hybrid DC Circuit Breaker

Dequan Wang, Minfu Liao *, Rufan Wang, Tenghui Li, Jun Qiu, Jinjin Li, Xiongying Duan * and Jiyan Zou

School of Electrical Engineering, Dalian University of Technology, Dalian 116024, China; wdq@mail.dlut.edu.cn (D.W.); Wangrufan@mail.dlut.edu.cn (R.W.); litenghui@mail.dlut.edu.cn (T.L.); qjun@mail.dlut.edu.cn (J.Q.); jinjinli@mail.dlut.edu.cn (J.L.); jyzou@dlut.edu.cn (J.Z.)

* Correspondence: LMF@dlut.edu.cn (M.L.); dxy@dlut.edu.cn (X.D.); Tel.: +86-198-1899-7875 (M.L.)

Received: 8 August 2020; Accepted: 12 September 2020; Published: 15 September 2020

Abstract: Vacuum arc commutation is an important process in natural-commutate hybrid direct current (DC) circuit breaker (NHCB) interruption, as the duration of vacuum arc commutation will directly affect the arcing time and interrupting time of NHCB. In this paper, the vacuum arc commutation model of NHCB was established by simplifying solid-state switch (SS) and vacuum arc voltage. Through theoretical analysis and experiments, the vacuum arc commutation characteristics of NHCB were studied. The mathematical formula of the effect of main parameters on the duration of vacuum arc commutation is obtained, and the changing law of the influence of the main parameters on the duration of the vacuum arc commutation is explored. The concept of vacuum arc commutation coefficient is proposed, and it is a key parameter that influences the vacuum arc commutation characteristics. The research on the characteristics of vacuum arc commutation can provide theoretical foundation for the structure and parameter optimization of NHCB and other equipment that uses vacuum arc commutation.

Keywords: hybrid dc circuit breaker; vacuum arc commutation; solid-state switch; vacuum arc voltage

1. Introduction

With the development of fully controlled power electronic devices (FPEDs), such as the gate turn-off thyristor (GTO), insulated gate bipolar transistor (IGBT), integrated gate commutated thyristor (IGCT), and injection-enhanced gate transistor (IEGT), the hybrid direct current (DC) circuit breaker based on power electronic technology has become an important technical solution and research hotspot for DC breaking [1–4]. This type of DC circuit breaker combines the advantages of solid-state switches (SSs) and mechanical switches (MSs), and shows the characteristics of a strong current carrying capacity, low on-state loss, fast breaking speed, and no need for cooling devices [5,6].

According to the difference of the current commutating method from MS to SS, hybrid DC circuit breakers based on power electronics technology are divided into forced-commutate hybrid DC circuit breakers (FHCBs) and natural-commutate hybrid DC circuit breakers (NHCBs) [7,8]. The FHCB is mostly used in the field of voltage above 10 kV [9,10], and one of the most representative FHCBs is the 320 kV/2.6 kA hybrid DC circuit breaker developed by ABB in 2011 [11]. NHCB is mostly used in the voltage field of 10kv and below, and there has been a lot of research on the NHCB [12]. Genji et al. [13] proposed the NHCB firstly in 1994, and SS of NHCB is mainly composed of GTO. Polman et al. [14] developed a 600V NHCB by fast MS and IGBTs in 2001, and it is a bi-directional switch. Luca Novello et al. [15] developed a prototype of 1 kV/10 kA NHCB by using IGCT's series and parallel technology in 2011. Lv gang et al. [16] studied the vacuum arc commutation characteristics of NHCB, and a 10 kV/3.5 kA NHDB prototype was designed and tested.

There have been some studies on the vacuum arc commutation characteristics of NHCB. The vacuum arc commutation model was constructed in [16,17], and the theoretical calculation formula of the rate of the current rise in the commutation branch is given. In [18], a theoretical analysis model is established, and the influence of adding impedance on vacuum arc commutation was studied. The authors of [19] simplified an equivalent treatment of the IGBT module in SS, and the criteria of the vacuum arc commutation were obtained. The authors of [15] studied the vacuum arc characteristics at different breaking distances and different currents. The authors of [20] studied the two current commutation measures for hybrid DC circuit breakers, and proposed a new current commutation measure. The authors of [21] studied the influence of internal and external factors on vacuum arc commutation, and reasonable parameters that can promote vacuum arc commutation were summarized.

The duration of vacuum arc commutation is a very important parameter, as it will directly affect the arcing time and interrupting time of NHCB, and shortening it will facilitate the interruption of NHCB. However, little work has focused on the duration of vacuum arc commutation, and, especially, there is a lack of mathematical formulas about it. In this regard, we conducted research on the vacuum arc commutation characteristics of NHCB, and focused on the influence of the main parameters on the duration of vacuum arc commutation. Section 2 establishes a simplified vacuum arc commutation model of NHCB, and theoretically derives the influence of the main parameters on the vacuum arc commutation characteristics. Sections 3 and 4 establish a vacuum arc commutation test circuit, and experimentally study the influence of the main parameters on the vacuum arc commutation characteristics. Section 5 first discusses the theoretical derivation and experimental results of the previous two sections, proposes the concept of the vacuum arc commutation coefficient, and then discusses the influence of the main parameters on the vacuum arc voltage. Finally, Section 6 concludes this paper.

2. Theoretical Analysis

2.1. Principle of NHCB

As shown in Figure 1, an NHCB mainly includes an MS, an SS and a metal-oxide varistor (MOV) [14]. The current limiting reactor is mainly used to limit the fault current. SS is mainly composed of FPEDs through series and parallel connection, and the most commonly used FPEDs are GTO, IGCT, IGBT, and IEGT. According to the literature [9], the interruption process of NHCB mainly includes vacuum arc commutation and the interruption process of SS. The interruption process of NHCB is shown in Figure 2. In this paper, we will focus on the vacuum arc commutation during NHCB interruption. I_m, I_s, and I_{mov} represent the currents of the MS, SS, and MOV, respectively.

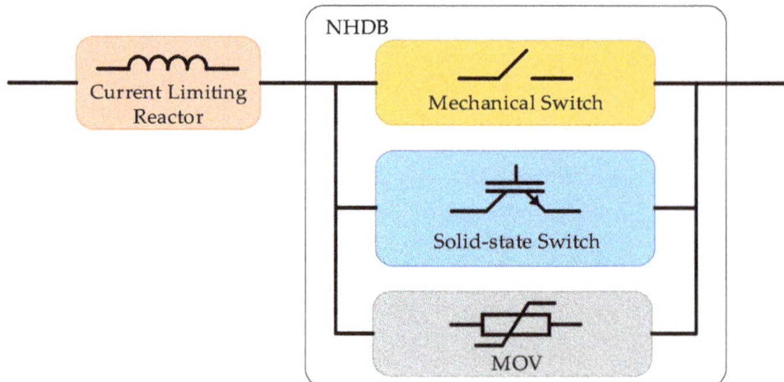

Figure 1. The schematic of a natural-commutate hybrid DC circuit breaker (NHCB).

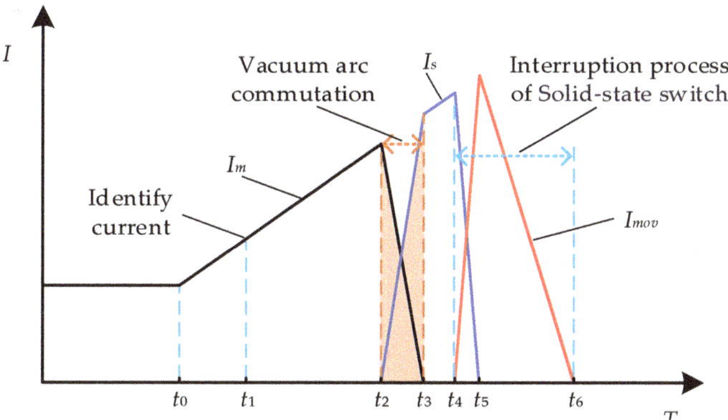

Figure 2. Interruption process of NHCB.

2.2. Vacuum Arc Commutation Model

The typical waveform of the vacuum arc commutation is shown in Figure 3 [18]. According to the research of the literature [16,17], the vacuum arc commutation model of NHCB is shown in Figure 4a. Then, according to the model shown in Figure 4a, the vacuum arc commutation should follow Formula (1). As shown in Formula (2), Formula (1) can be further rewritten as the relational expression about the current commutating speed $\mathrm{d}I_c/\mathrm{d}t_c$. U_{arc} represents the vacuum arc voltage. U_{se} represents the on-state voltage of the SS at the end of vacuum arc commutation. L_s represents the stray inductance of the SS. I_n represents the currents of NHCB. I_{si} represents the current of the SS at the initial moment of vacuum arc commutation. I_{se} represents the current of the SS at the end of vacuum arc commutation. I_c is called the commutation current, which represents the current commutated to the SS during the vacuum arc commutation. I_{ce} is called the final commutation current, which represents the current commutated to the SS from the beginning to the end of the vacuum arc commutation. t_c represents the time it takes for the commutation current to increase from zero to I_c. T_c represents the duration of vacuum arc commutation.

$$U_{arc} = L_s \frac{\mathrm{d}I_c}{\mathrm{d}t_c} + U_s \; 0 \le I_c \le I_{ce}, \tag{1}$$

$$\frac{\mathrm{d}I_c}{\mathrm{d}t_c} = \frac{U_{arc} - U_s}{L_s} \; 0 \le I_c \le I_{ce}. \tag{2}$$

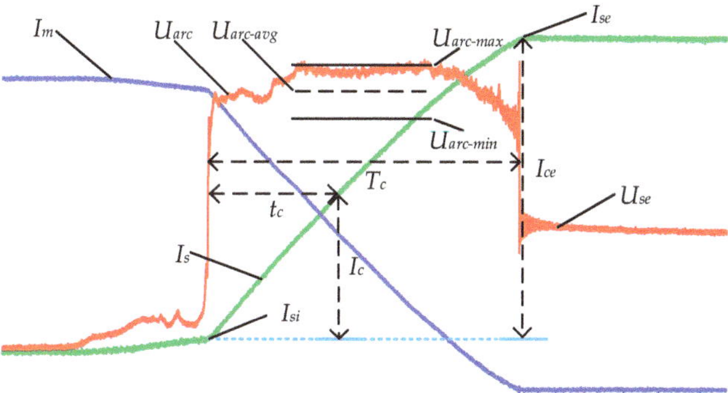

Figure 3. Example of typical vacuum arc commutation.

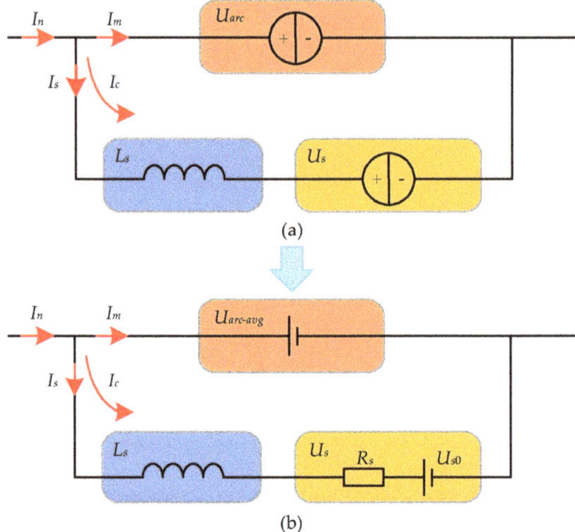

Figure 4. The vacuum arc commutation model of NHCB (**a**) before and (**b**) after simplification.

For FPEDs, such as IGBT, IGCT, and IEGT, the relationship between their on-state voltage U_{CE} and collector current I is shown in Formula (3) [22]. As shown in Figure 5, the value of their on-state voltage U_{CE} is basically linear with the collector current I. Therefore, as shown in Figure 5, the relationship between U_{CE} and I can be linearly processed, and the relationship between U_{CE} and I is obtained as shown in Formula (4). Because SSs are made of FPEDs through series and parallel connection, the relationship between U_s and I_s can also be obtained as shown in Formula (5). U_{CE0} represents the on-state voltage of the FPEDs when the collector current is zero. r represents the on-state resistance of the FPEDs. U_0 represents the approximate on-state voltage of the FPEDs when the collector current is zero. R represents the approximate on-state resistance of the FPEDs. U_{s0} represents the approximate on-state voltage of the SS when the I_s is zero:

$$U_{CE} = U_{CE0} + rI, \tag{3}$$

$$U_{CE} = U_0 + RI, \tag{4}$$

$$U_s = U_{s0} + R_s I_s. \tag{5}$$

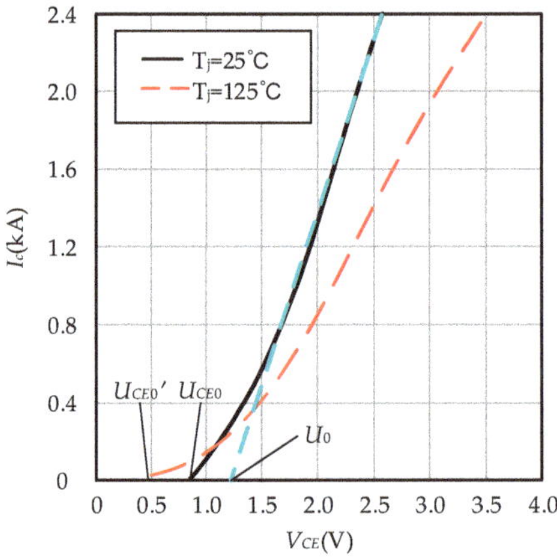

Figure 5. Collector-emitter saturation voltage characteristics of FZ1200R17HP4_B2 [23].

As shown in Figure 3, because the change in the vacuum arc voltage is relatively small during the vacuum arc commutation [24], $U_{arc\text{-}avg}$ can be taken as the vacuum arc voltage value, and $U_{arc\text{-}avg}$ is the average value of the maximum value $U_{arc\text{-}max}$ and the minimum value $U_{arc\text{-}min}$ of the vacuum arc voltage. According to Formula (5), the model shown in Figure 4a can be rewritten as shown in Figure 4b. Then, according to the model shown in Figure 4b, the vacuum arc commutation should follow Formula (6). It is easy to know that $I_s = I_{si} + I_c$, and according to Formula (5), the system of equations shown in Formula (7) can be easily obtained. U_{si} represents the on-state voltage of the SS at the initial moment of vacuum arc commutation:

$$\frac{\mathrm{d}I_c}{\mathrm{d}t_c} = \frac{U_{arc\text{-}avg} - U_s}{L_s}\ 0 \le I_c \le I_{ce}, \tag{6}$$

$$\begin{cases} U_{si} = U_{s0} + R_s I_{si} \\ U_s = U_{si} + R_s I_c\ 0 \le I_c \le I_{ce} \end{cases}. \tag{7}$$

After solving Formulas (6) and (7), the relationship expression of t_c about I_c is obtained as shown in Formula (8). It is easy to know that when $I_c = I_{ce}$, $t_c = T_c$, then according to Formula (8), the relationship between T_c and I_{ce} can be obtained as shown in Formula (9). From Formula (9), we can see that L_s, R_s, I_{ce}, $U_{arc\text{-}avg}$, and U_{si} are the main parameters that affect T_c. Meanwhile, according to Formula (6), if $\mathrm{d}I_c/\mathrm{d}t_c \le 0$, the current will stop commuting to the SS branch, so $\mathrm{d}I_c/\mathrm{d}t_c > 0$ should always be ensured during the vacuum arc commutation. By solving $\mathrm{d}I_c/\mathrm{d}t_c > 0$, Formula (10) can be obtained:

$$t_c = -\frac{L_s}{R_s}\ln\left(\frac{U_{arc\text{-}avg} - U_{si} - R_s I_c}{U_{arc\text{-}avg} - U_{si}}\right) 0 \le I_c \le I_{ce}, \tag{8}$$

$$T_c = -\frac{L_s}{R_s}\ln\left(\frac{U_{arc\text{-}avg} - U_{si} - R_s I_{ce}}{U_{arc\text{-}avg} - U_{si}}\right), \tag{9}$$

$$U_{arc\text{-}avg} - U_{si} - R_s I_{ce} > 0. \tag{10}$$

2.3. Influence of Main Parameters on Vacuum Arc Commutation Characteristics

In this section, the effects of parameters, such as L_s, R_s, I_{ce}, $U_{arc\text{-}avg}$, and U_{si}, on T_c are further studied. Firstly, it is easy to know from Formula (9), under the condition that other parameters remain unchanged, T_c has a linear relationship with L_s.

Assuming that the parameters L_s and R_s remain unchanged and L_s/R_s = b, let $k = (U_{arc\text{-}avg} - U_{si})/R_s I_{ce}$, then as shown in Formula (11), Formula (9) can be further rewritten as a function about k, and it is easy to know from Formula (10) that $k > 1$:

$$T_c = -\mathrm{b}\cdot\ln\left(1 - \frac{1}{k}\right) k > 1. \tag{11}$$

It is easy to know that changing the parameters $U_{arc\text{-}avg}$, U_{si}, and I_{ce} is equivalent to changing the parameter k, and the relationship curve between T_c and k is plotted according to Formula (11), as shown in Figure 6. It can be seen from Figure 6, T_c decreases with increasing k, but the rate of decrease becomes slower and slower and even tends to be constant. k is the ratio of the parameters $U_{arc\text{-}avg} - U_{si}$ and $R_s I_{ce}$.

Assuming that the parameters L_s, I_{ce}, $U_{arc\text{-}avg}$ and U_{si} remain unchanged and L_s = a, the variation of T_c with R_s is explored. Let the value of R_s at the initial time be a fixed value R_{s0}. Correspondingly, $k = (U_{arc\text{-}avg} - U_{si})/R_{s0} I_{ce} = k_0$; let the changed $R_s = xR_{s0}$, and correspondingly, $k = (U_{arc\text{-}avg} - U_{si})/xR_{s0} I_{ce} = k_0/x$. Then, as shown in Formula (12), Formula (9) can be further rewritten as a function about x, and it is easy to know from Formula (10) that $x < k_0$:

$$T_c = -\frac{\mathrm{a}}{x}\ln\left(1 - \frac{x}{k_0}\right) x < k_0. \tag{12}$$

When k_0 takes the values of 4 and 8, respectively, as shown in Figure 7, the variation curves of T_c and k are plotted with x according to Formula (12). It can be seen from Figure 7, T_c increases with increasing x, and the rate of increase becomes faster and faster; k decreases with increasing x, but the rate of decrease becomes slower and slower. The change curve of T_c with k is plotted in Figure 7 as shown in Figure 8. As it can be seen from Figure 8, T_c decreases with increasing k, but the rate of decrease becomes slower and slower and even tends to be constant. k_0 is the ratio of the parameters $U_{arc\text{-}avg} - U_{si}$ and $R_{s0}I_{ce}$, and R_{s0} is the initial value of R_s.

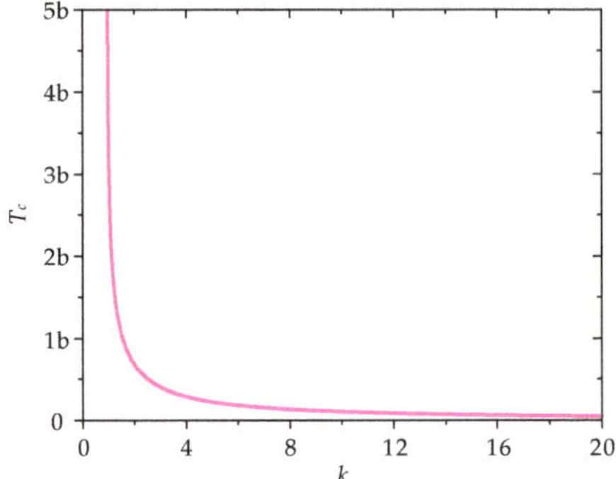

Figure 6. The relationship curve between t_c and k.

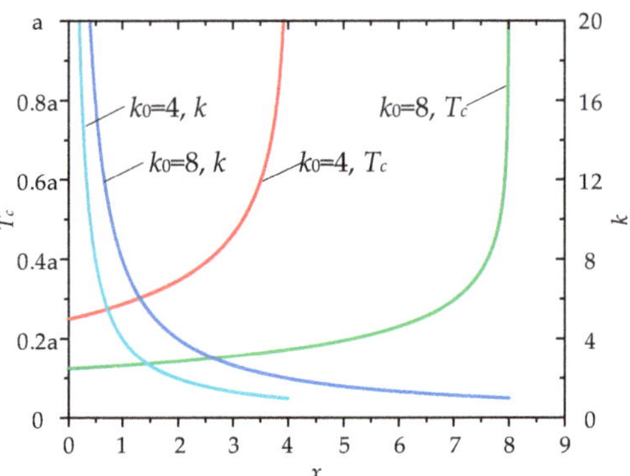

Figure 7. The variation curves of T_c and k with x.

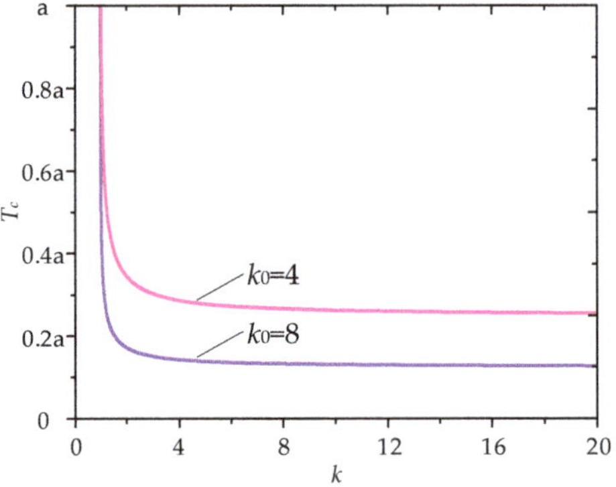

Figure 8. The relationship curve between T_c and k in Figure 7.

It can be seen from Figures 6 and 8 that when k is small, T_c changes greatly with the increase of k, and when k is large, T_c changes little or even tends to be constant with the increase of k; therefore, the larger the value of k, the more effective it is to reduce the value of T_c.

3. Test Configuration

As shown in Figure 9, the test platform is mainly composed of an LC oscillation circuit, NHCB, controller, and data acquisition system. The LC oscillation circuit is composed of a capacitor C_i and an inductance L_i, and is mainly used to generate the large current required for the experiment. K_m is the experimental control switch, responsible for introducing current into NHCB. K_v represents the MS of NHCB. The SS of NHCB is mainly composed of impedance Z_a and the IGBT module, and the impedance Z_a is formed by the inductance L_a and the resistance R_a in series. The on-state resistance and stray inductance of the SS are changed by changing R_a and L_a, respectively, to explore the effects of on-state resistance and stray inductance on vacuum arc commutation characteristics of NHCB. The IGBT selected in this experiment is CM1000HA-24H, the collector current of a single CM1000HA-24H is up to 2 kA, and two IGBTs are connected in parallel to form an IGBT module. C_{t1} and C_{t2} are current sensors, which are used to collect the currents of the MS and SS branches, respectively. P_t is a voltage sensor used to collect the vacuum arc voltage. The controller is used to control the operation sequence of the entire experiment. The data acquisition system is used for data collection and display.

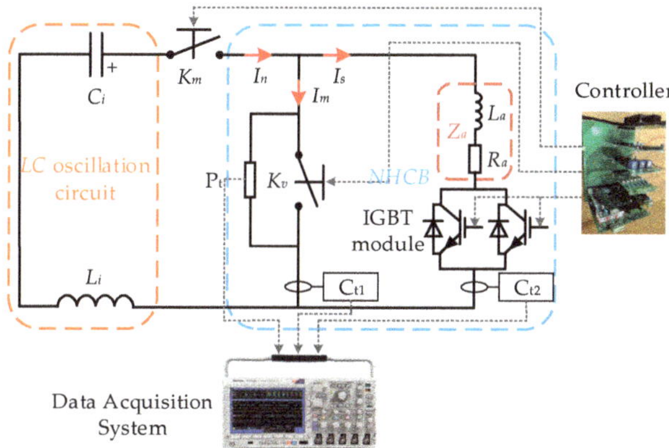

Figure 9. Vacuum arc commutation test platform.

4. Experimental Results

4.1. Influence of Stray Inductance of the SS

Under the condition that R_a is 5 mΩ, and I_{se} is approximately 0.5, 1, and 1.5 kA, respectively, we obtained the waveforms of vacuum arc commutation with different stray inductances. When the I_{se} is approximately 1 kA, the waveforms when the added stray inductance L_a is 0 and 2 μH are shown in Figure 10a,b, respectively. As shown in Figure 10, when the added stray inductance L_a increases from 0 to 2 μH, T_c increases from 66 to 346 μs, and $U_{arc\text{-}avg}$ increases from 11.7 to 15.8 V. Both T_c and $U_{arc\text{-}avg}$ increase with the increase of L_a.

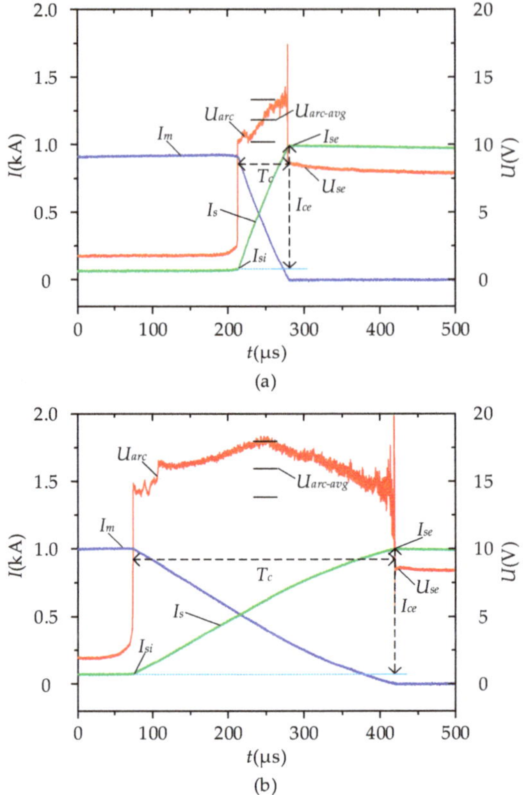

Figure 10. Waveforms of vacuum arc commutation when the added stray inductance L_a is (**a**) 0 and (**b**) 2 μH.

When the added stray inductance L_a ranges from 0 to 4 μH, the relationship curve between T_c and $U_{arc\text{-}avg}$ and L_a is shown in Figure 11. As shown in Figure 11, the relationship between T_c and L_a is approximately linear, which conforms to the law shown in Formula (9), and $U_{arc\text{-}avg}$ also increases with the increase of L_a, but the magnitude of the increase becomes smaller and smaller.

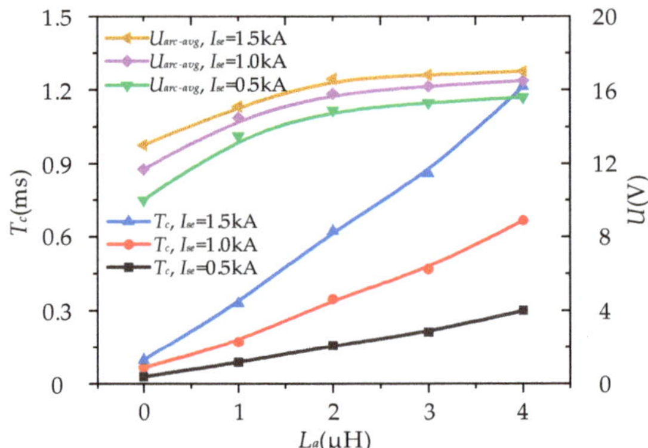

Figure 11. Curve of T_c and $U_{arc\text{-}avg}$ with different stray inductances.

4.2. Influence of the Final Commutation Current

Under the condition that R_a is 2 mΩ, and L_a is 0 μH, we obtained the waveforms of vacuum arc commutation with different commutation currents I_{ce}. The waveforms when the commutation current I_{ce} is 0.28 and 2.68 kA are shown in Figure 12a,b, respectively. As shown in Figure 12, when the commutation current I_{ce} increases from 0.28 to 2.68 kA, T_c increases from 13 to 195 μs, $U_{arc\text{-}avg}$ increases from 9.2 to 12.9 V, U_{se} increases from 2.7 to 11.8 V, and I_{si} increases from 0.02 to 0.48 kA. All of T_c, $U_{arc\text{-}avg}$, U_{se}, and I_{si} increase with the increase of I_{ce}.

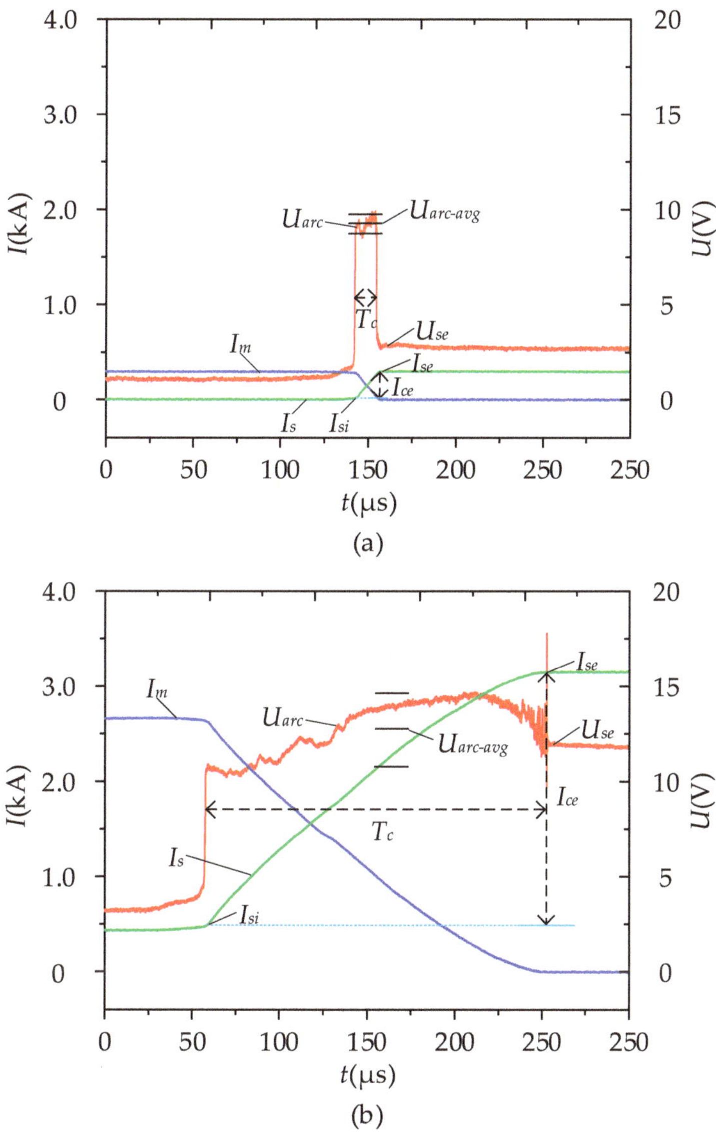

Figure 12. Waveforms of vacuum arc commutation when the final commutation current I_{ce} is (**a**) 0.28 and (**b**) 2.68 kA.

As shown in Figure 12, it is easy to know that changing I_{ce} is achieved by changing I_{se}. When the I_{se} ranges from 3.0 to 3.16 kA, the relationship curve between $U_{arc\text{-}avg}$, U_{se}, I_{ce}, I_{si}, and I_{se} is shown in Figure 13a. As shown in Figure 13a, all of $U_{arc\text{-}avg}$, U_{se}, I_{ce}, and I_{si} increase with the increase of I_{se}, and according to Figure 5 and Formula (7), as shown in Formula (13), the linear expressions between U_{se} and I_{se} and U_{si} and I_{si} can be obtained, respectively, and then we can get the calculation formula of k about $U_{arc\text{-}avg}$, U_{se} and U_{si}:

$$\begin{cases} U_{se} = U_{s0} + R_s I_{se} = 2.0 + 3I_{se} \\ U_{si} = U_{s0} + R_s I_{si} = 2.0 + 3I_{si} \\ k = \frac{U_{arc\text{-}avg} - U_{si}}{R_s I_{ce}} = \frac{U_{arc\text{-}avg} - U_{si}}{U_{se} - U_{si}} \end{cases} . \quad (13)$$

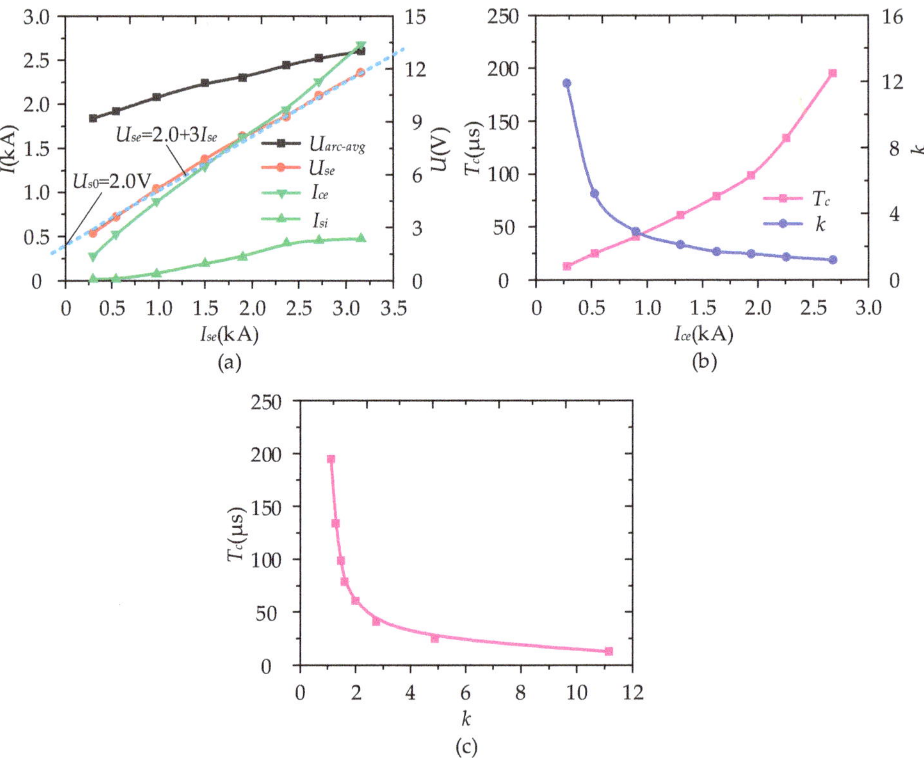

Figure 13. (**a**) The relationship curve between $U_{arc\text{-}avg}$, U_{se}, I_{ce}, I_{si}, and I_{se}; (**b**) Curve of T_c and k with different final commutation currents; (**c**) The relationship curve between T_c and k in (**b**).

Therefore, according to Figure 13a and Formula (13), the k value corresponding to different I_{ce} values can be calculated, and the relationship curve between k and T_c and I_{ce} is shown in Figure 13b. As shown in Figure 13b, k decreases with the increase of I_{ce}, and the rate of decrease becomes slower and slower; T_c increases with the increase of I_{ce}, and the rate of increase becomes faster and faster. The relationship curve between T_c and k was plotted according to Figure 13b, as shown in Figure 13c. It can be seen from Figure 13c that T_c decreases with increasing k, and the rate of decrease becomes slower and slower; the change law shown in Figure 13c basically coincides with the change law shown in Formula (11) and Figure 6.

4.3. Influence of the On-State Resistance of the SS

Under the condition that L_a is 0 μH, and I_{ce} is approximately 1 and 2 kA, respectively, we obtained the waveforms of vacuum arc commutation with different on-state resistances. When the I_{ce} is approximately 1 kA, the waveforms when the added on-state resistances R_a is 0 and 9 mΩ are shown in Figure 14a,b, respectively. Because when the added on-resistance is different, the value of I_{si} will also be different; so, to eliminate the impact of different I_{si}, as shown in Figure 14, the IGBT module is turned on after the vacuum arc is generated, and in this case, I_{si} = 0 and I_{ce} = I_{se}. As shown in Figure 14, when the added on-state resistances R_a increases from 0 to 9 mΩ, T_c increases from 34 to 99 μs, $U_{arc\text{-}avg}$ increases from 9.5 to 12.8 V, and U_{se} increases from 3 to 12.1 V. All of T_c, $U_{arc\text{-}avg}$, and U_{se} increase with the increase of R_a. Meanwhile, it can be seen in Figure 14 that when the IGBT module is turned on, the vacuum arc voltage drops.

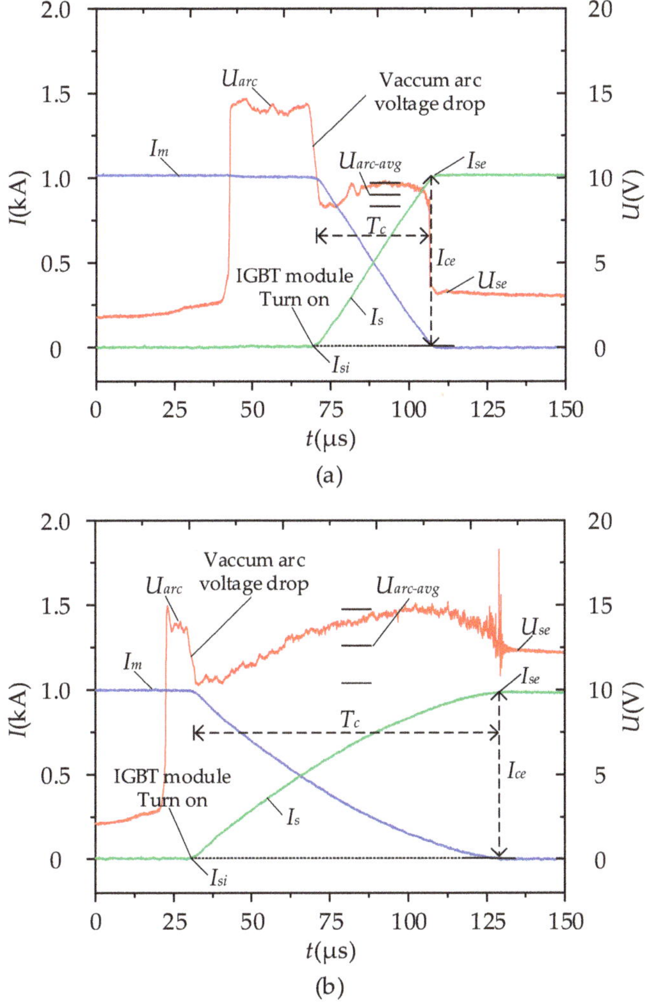

Figure 14. Waveforms of vacuum arc commutation when the added on-state resistances R_a is (**a**) 0 and (**b**) 9 mΩ.

When I_{ce} is 1 and 2 kA, respectively, the relationship between $U_{arc\text{-}avg}$, U_{se}, and R_a is shown in Figure 15a. It can be seen from Figure 15a that whether I_{ce} is 1 or 2 kA, both $U_{arc\text{-}avg}$ and U_{se} grow approximately linearly with R_a. As it is known from the previous section that $U_{s0} = 2.0$, and $I_{si} = 0$, we can easily get $U_{si} = U_{s0} + I_{si}R_s = 2.0$. Therefore, according to Figure 15a and Formula (13), the k value corresponding to different R_a values can be calculated, and the relationship between k and T_c and R_a is shown in Figure 15b. It can be seen from Figure 15b that whether I_{ce} is 1 or 2 kA, k decreases with the increase of R_a, and the rate of decrease becomes slower and slower; T_c increases with the increase of R_a, and the rate of increase becomes faster and faster; the change law shown in Figure 15b basically coincides with the change law shown in Figure 7. Meanwhile, when $R_a = 0$, the corresponding k_0 value is 8.2 when the I_{ce} is 1 kA, and the corresponding k_0 value is 4.3 when the I_{ce} is 2 kA. The relationship curve between T_c and k was plotted according to Figure 15b, as shown in Figure 15c. It can be seen from Figure 15c that T_c decreases with increasing k, and the rate of decrease becomes slower and slower and even tends to be constant; the change law shown in Figure 15c basically coincides with the change law shown in Formula (12) and Figure 8.

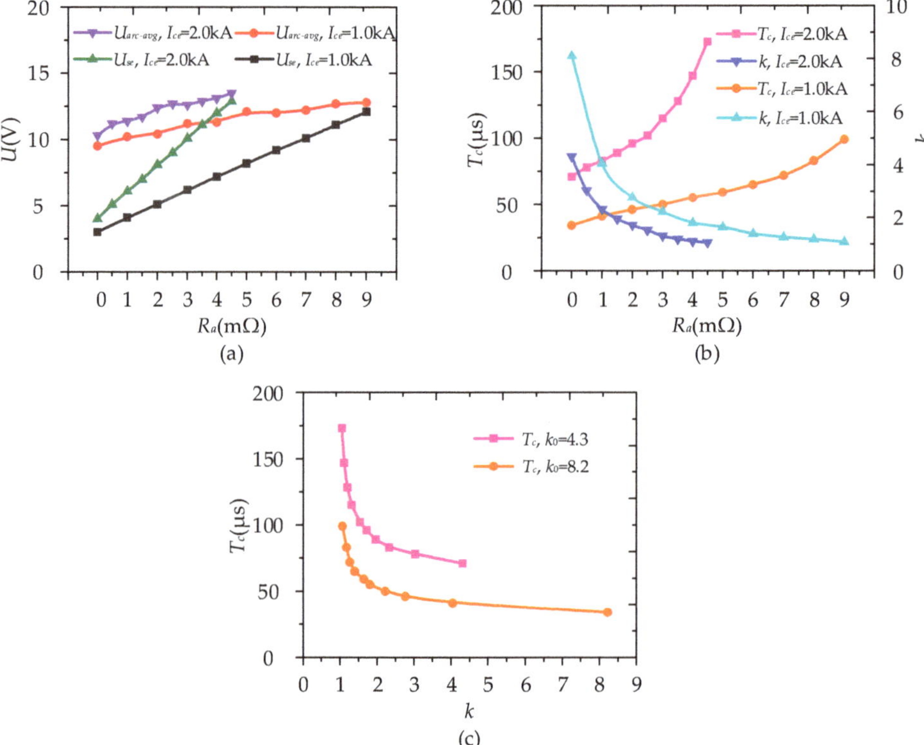

Figure 15. (**a**) Curve of $U_{arc\text{-}avg}$ and U_{se} with different on-state resistances; (**b**) Curve of T_c and k with different on-state resistances; (**c**) The relationship curve between T_c and k in (**b**).

5. Discussion

5.1. Vacuum Arc Commutation Coefficient

Compared to the theoretical analysis and experiments above, it is easy to find that the experimental results can better match the theoretical Formulas (9)–(12). This shows that Formulas (9)–(12) can better reflect the influence of the main parameters, such as the vacuum arc voltage U_{arc}, final commutation current I_{ce}, on-state resistance R_s, and stray inductance L_s, on the vacuum arc commutation characteristics. Meanwhile, according to Formulas (11) and (12) and Figures 6, 8, 13c and 15c, the ratio k of $U_{arc\text{-}avg}$-U_{si} to R_sI_{ce} is a key parameter that affects T_c; and according to Formula (10), it is necessary to ensure that $k > 1$ so that the current can be commutated to the SS branch. Overall, k is a very important parameter during the vacuum arc commutation. Therefore, this paper defines k as the vacuum arc commutation coefficient.

According to Figures 6, 8, 13c and 15c, when k is small, T_c will increase significantly. Therefore, in order to reduce T_c, the value of $U_{arc\text{-}avg} - U_{si}$ should be increased or the value of R_sI_{ce} should be decreased. The main method of increasing the value of $U_{arc\text{-}avg} - U_{si}$ is to increase the vacuum arc voltage. Applying an external magnetic field and connecting multiple MSs in series are currently common methods of increasing the arc voltage [25,26]. The main method to reduce the value of R_sI_{ce} is to reduce the on-resistance R_s and final commutation current I_{ce}, but reducing I_{ce} will reduce the interrupting ability of NHCB. The method to reduce the on-resistance R_s is to select FPEDs with low on-resistance or to increase the number of parallel FPEDs [27,28]. Meanwhile, it can be seen from Formula (9) and Figure 11 that reducing the stray inductance L_s in the SS is also an effective method to reduce T_c.

5.2. *Influence of the Main Parameters on the Vacuum Arc Voltage*

It can be seen from the experiment that when the parameters L_s, I_{ce}, and R_s increase, not only T_c increases but the vacuum arc voltage also increases, which is consistent with the positive feedback characteristics of the vacuum arc voltage. This can explain the sudden drop of the vacuum arc voltage in Figure 14, before the IGBT module is turned on, the vacuum arc commutation is very difficult, so the vacuum arc voltage at this time is higher. When the IGBT module is turned on, the difficulty of vacuum arc commutation decreases rapidly, so the vacuum arc voltage also decreases accordingly. Meanwhile, to better describe the influence of parameters L_s, I_{ce}, and R_s on T_c, in the future work, the influence of the parameters L_s, I_{ce}, and R_s on the vacuum arc voltage should be considered in Formula (9).

Although this paper focuses on NHCB, it is also applicable to other devices that use vacuum arc commutation, such as fault current limiters [29,30]. In the future work, further research will be conducted on other devices that use vacuum arc commutation.

6. Conclusions

In this paper, the vacuum arc commutation model of NHCB was established by simplifying SS and vacuum arc voltage, and the vacuum arc commutation characteristics of NHCB were studied through theoretical analysis and experiments. The mathematical expression of the relationship between the main parameters of the vacuum arc voltage, on-state resistance, and stray inductance and the duration of vacuum arc commutation was obtained.

The concept of the vacuum arc commutation coefficient was proposed based on the mathematical expressions and experiments, and it is a key parameter that influences the vacuum arc commutation characteristics. The value of the vacuum arc commutation coefficient should not be small; when it is small, the duration of vacuum arc commutation will increase significantly. Further, it is not the bigger the better, as when it is bigger, the duration of vacuum arc commutation will decrease slowly. Meanwhile, it must always be ensured that the value of the vacuum arc commutation coefficient is greater than one, so that the current can be commutated to the SS branch.

In the process of vacuum arc commutation, when the difficulty of vacuum arc commutation increases, the vacuum arc voltage will also increase, and from the extremely difficult vacuum arc commutation to the easier vacuum arc commutation, the vacuum arc voltage will also be reduced immediately. This paper can provide the theoretical foundation for the structure and parameter optimization of NHCB and other devices using vacuum arc commutation.

Author Contributions: D.W. and M.L. contributed to the research idea. D.W., X.D. and J.Z. designed the experiments. D.W. worked on the theoretical analysis, data analysis and drafted the manuscript. R.W. and T.L. helped to do the experiments. J.Q. and J.L. worked on editing and revising of the manuscript. All authors have read and agreed to the published version of the manuscript.

Funding: This research was funded in part by the National Natural Science Foundation of China (No. 51777025) and in part by the Fundamental Research Funds for the Central Universities (No. DUT19ZD219).

Conflicts of Interest: The authors declare no conflict of interest.

References

1. Chen, Z.; Yu, Z.; Zhang, X.; Wei, T.; Lyu, G.; Qu, L.; Huang, Y.; Zeng, R. Analysis and Experiments for IGBT, IEGT, and IGCT in Hybrid DC Circuit Breaker. *IEEE Trans. Ind. Electron.* **2018**, *65*, 2883–2892. [CrossRef]
2. Martin, B.; Jens, W.; Steffen, B. Comparison of IGCT and IGBT for the use in the modular multilevel converter for HVDC applications. In Proceedings of the IEEE 9th International Multi-Conference on Systems, Signals and Devices, Chemnitz, Germany, 20–23 March 2012.
3. Bernet, S.; Teichmann, R.; Zuckerberger, A.; Steimer, P. Comparison of high-power IGBT's and hard-driven GTO's for high-power inverters. *IEEE Trans. Ind. Appl.* **1999**, *35*, 487–497. [CrossRef]
4. Zhou, Y.; Huang, Y.; Pang, J.; Wang, K. Remaining useful life prediction for supercapacitor based on long short-term memory neural network. *J. Power Sources* **2019**, *440*, 227149. [CrossRef]
5. Geider, P.V.; Ferreira, J.A. Zero volt switching hybrid DC circuit breakers. In Proceedings of the IEEE Industry Applications Society Annual Meeting, Rome, Italy, 8–12 October 2000; Volume 5, pp. 2923–2927.
6. Ahmad, M.; Wang, Z. A Hybrid DC Circuit Breaker with Fault-Current-Limiting Capability for VSC-HVDC Transmission System. *Energies* **2019**, *12*, 2388. [CrossRef]
7. Swati, R.; Devenkumar, K.; Chragkumar, D.; Subrata, P. Development of a Prototype Hybrid DC Circuit Breaker for Superconducting Magnets Quench Protection. *IEEE Trans. Appl. Supercond.* **2014**, *24*, 4702006.
8. Gu, C.; Wheeler, P.; Castellazzi, A.; Watson, A.J.; Effah, F. Semiconductor Devices in Solid-State/Hybrid Circuit Breakers: Current Status and Future Trends. *Energies* **2017**, *10*, 495. [CrossRef]
9. Zhang, X.; Yu, Z.; Zhao, B.; Chen, Z.; Lyu, G.; Huang, Y.; Zeng, R. A Novel Mixture Solid-State Switch Based on IGCT With High Capacity and IGBT With High Turn- OFF Ability for Hybrid DC Breakers. *IEEE Trans. Ind. Electron.* **2020**, *67*, 4485–4495. [CrossRef]
10. Nadeem, M.H.; Zheng, X.; Tai, N.; Gul, M. Identification and Isolation of Faults in Multi-terminal High Voltage DC Networks with Hybrid Circuit Breakers. *Energies* **2018**, *11*, 1086. [CrossRef]
11. Callavik, M.; Blomberg, A.; Häfner, J.; Jacobson, B. Break-through!: ABB's hybrid HVDC breaker, an innovation breakthrough enabling reliable HVDC grids. *ABB Rev.* **2013**, *2*, 7–13.
12. Meyer, J.M.; Rufer, A. A DC hybrid circuit breaker with ultra-fast contact opening and integrated gate-commutated thyristors (IGCT). *IEEE Trans. Power Del.* **2006**, *21*, 646–651. [CrossRef]
13. Genji, T.; Nakamura, O.; Isozaki, M.; Yamada, M.; Morita, T.; Kaneda, M. 400V class high-speed current limiting circuit breaker for electric power system. *IEEE Trans. Power Del.* **1994**, *9*, 1428–1435. [CrossRef]
14. Polman, H.; Ferreira, J.; Kaanders, M.; Evenblij, B.; Gelder, P. Design of a bi-directional 600V/6kA ZVS hybrid DC switch using IGBTs. *Ind. Appl. Conf.* **2001**, *2*, 1052–1059.
15. Novello, L.; Baldo, F.; Ferro, A.; Maistrello, A.; Gaio, E. Development and Testing of a 10-kA Hybrid Mechanical-Static DC Circuit Breaker. *IEEE Trans. Appl. Supercond.* **2011**, *21*, 3621–3627. [CrossRef]
16. Lv, G.; Zeng, R.; Huang, Y.; Chen, Z. Researches on 10 kV DC hybrid circuit breaker based on IGCT series. *Chin. Soc. For Elec. Eng.* **2017**, *37*, 1012–1020.
17. Roodenburg, B.; Taffone, A.; Gilardi, E.; Tenconi, S.M.; Evenblij, B.H.; Kaanders, M.A.M. Combined ZVS—ZCS topology for high-current direct current hybrid switches: Design aspects and first measurements. *IET Electr. Power Appl.* **2007**, *1*, 183–192. [CrossRef]
18. Liao, M.; Wang, D.; Leng, T.; Duan, X.; Wang, R. Effect of Adding Impedance for Current Balancing on the Vacuum Arc Commutation Characteristics of DC Microgrid Hybrid Circuit Breakers. *IEEJ Trans.* **2020**, *15*, 676–683. [CrossRef]
19. Liao, M.; Huang, J.; Ge, G.; Duan, X.; Huang, Z.; Zou, J. Vacuum arc commutation characteristics of the DC microgrid hybrid circuit breakers. *IEEE Trans. Plasma Sci.* **2017**, *45*, 2172–2178. [CrossRef]

20. Wen, W.; Huang, Y.; Sun, Y.; Wu, J.; Mohmmad, A.; Liu, W. Research on Current Commutation Measures for Hybrid DC Circuit Breakers. *IEEE Trans. Power Del.* **2016**, *31*, 1456–1463. [CrossRef]
21. Dong, E.; Zhang, L.; Wang, S.; Qin, T.; Zou, J.; Liu, J. Research on Internal and External Factors Affecting Current Commutation in Vacuum. In Proceedings of the 28th International Symposium on Discharges and Electrical Insulation in Vacuum (ISDEIV 2018), Greifswald, Germany, 23–28 September 2018; Volume 1, pp. 279–282.
22. Felipe, F.; Rodrigo, A.; Steffen, B. Comparison of 4.5-kV press-pack IGBTs and IGCTs for medium-voltage converters. *IEEE Trans. Ind. Electron.* **2013**, *60*, 440–449.
23. Auerbach, F.; Bauer, M.; Gottert, J.; Hierholzer, M.; Porst, A.; Reznik, D.; Schulze, H.; Schulze, T.; Spanke, R. 6.5 kV IGBT-modules. In Proceedings of the IEEE industry applications society annual meeting, Phoenix, AZ, USA, 3–7 October 1999; Volume 3, pp. 1770–1774.
24. Alferov, D.F.; Ivanov, V.P.; Sidorov, V.A. DC vacuum arc in a Performance of a nonuniform axisymmetric magnetic field. *High Temp.* **2006**, *44*, 342–355. [CrossRef]
25. Alferov, D.F.; Ivanov, V.P.; Sidorov, V.A. Characteristics of DC vacuum arc in the transverse axially symmetric magnetic field. *IEEE Trans. Plasma Sci.* **2003**, *31*, 918–922. [CrossRef]
26. Zabello, K.K.; Barinov, Y.A.; Chaly, A.M.; Logatchev, A.A.; Shkolnik, S.M. Experimental study of cathode spot motion and burning voltage of low-current vacuum arc in magnetic field. *IEEE Trans. Plasma Sci.* **2005**, *33*, 1553–1559. [CrossRef]
27. Letor, R. Static and dynamic behavior of paralleled IGBTs. *IEEE Trans. Ind. Appl.* **1992**, *28*, 395–402. [CrossRef]
28. Shammas, N.Y.A.; Withanage, R.; Chamund, D. Optimisation of the number of IGBT devices in a series–parallel string. *Microelectron. J.* **2008**, *39*, 899–907. [CrossRef]
29. Pedrow, P.D.; Burrage, L.M.; Shohet, J.L. Performance of a Vacuum Arc Commutating Switch for a Fault-Current Limiter. *IEEE Power Energy Mag.* **1983**, *5*, 1269–1277.
30. Khan, U.F.; Lee, J.; Amir, F.; Lee, B. A Novel Model of HVDC Hybrid-Type Superconducting Circuit Breaker and Its Performance Analysis for Limiting and Breaking DC Fault Currents. *IEEE Trans. Appl. Supercond.* **2015**, *25*, 1–9. [CrossRef]

Article

Experimental Investigation of the Prestrike Characteristics of a Double-Break Vacuum Circuit Breaker under DC Voltages

Yun Geng [1], Xiaofei Yao [1,*], Jinlong Dong [1,*], Xue Liu [1], Yingsan Geng [1], Zhiyuan Liu [1], Jing Peng [2] and Ke Wang [2]

1 State Key Laboratory of Electrical Insulation and Power Equipment, Xi'an Jiaotong University, No.28 Xianning West Road, Xi'an 710049, China; gengcloud@163.com (Y.G.); xuel96723@163.com (X.L.); ysgeng@xjtu.edu.cn (Y.G.); liuzy@xjtu.edu.cn (Z.L.)

2 Electric Power Research Institute, Yunnan Power Grid Co. Ltd., No.105 Yunda West Road, Kunming 650217, China; pengjingyunnan@126.com (J.P.); wangkeyunnan@126.com (K.W.)

* Correspondence: yaoxf85@mail.xjtu.edu.cn (X.Y.); djl.1989@stu.xjtu.edu.cn (J.D.); Tel.: +86-029-82663773 (J.D.)

Received: 1 May 2020; Accepted: 16 June 2020; Published: 20 June 2020

Abstract: The prestrike phenomenon in vacuum circuit breakers (VCBs) is interesting but complicated. Previous studies mainly focus on the prestrike phenomenon in single-break VCBs. However, experimental work on prestrike characteristics of double-break VCBs cannot be found in literature. This paper aims to experimentally determine the probabilistic characteristics of prestrike gaps in a double-break VCB consisting of two commercial vacuum interrupters (VIs) in series under direct current (DC) voltages. As a benchmark, single-break prestrike gaps were measured by short-circuiting one of the VIs in a double break. The experimental results show that the 50% prestrike gap d_{50} of each VI in a double break, which is calculated with the complementary Weibull distribution, was significantly reduced by 25% to 72.7% compared with that in a single break. Due to the voltage-sharing effect in the double-break VCB, scatters in prestrike gaps of each VI in a double break was smaller than that in a single break. However, without the sharing-voltage effect, d_{50} of the low-voltage side in the double break was 65% higher than that of the same VI in the single break, which could be caused by the asynchronous property of mechanical actuators, the difference of the inherent prestrike characteristics of each VI and the unequal voltage-sharing ratio of VIs.

Keywords: vacuum circuit breaker; double break; prestrike characteristics; vacuum interrupter; prestrike gap

---

1. Introduction

Vacuum circuit breakers (VCBs) are widely used in medium voltage power systems while SF_6 gas circuit breakers are still the prevailing technology in high- and extra high-voltage networks. However, SF_6 gas has been specified as a strong greenhouse gas since 1997 in the Kyoto Protocol and its emission has been strictly restricted. Therefore, it has been of increasing interest to develop VCBs to higher voltage levels due to their advantages such as being maintenance free, having a long mechanical life, excellent dielectric strength, high interruption capability, and environment-friendly characteristics [1].

The major difficulty of extending the application of VCBs to higher voltage levels arises from the physical disadvantage of a vacuum in terms of the dielectric characteristics, i.e., the non-linear relationship between the dielectric strength and contact gap length [2]. To be precise, the breakdown voltage is nearly linear to the vacuum gap length within about 5 mm [3]. However, for a larger gap, the dielectric strength shows a strong full voltage effect. An appropriate way to develop high voltage level VCBs is to use the double-or multi-break VCBs consisting of two or more vacuum interrupters (VIs) in series, which takes full advantages of the excellent dielectric strength and high interruption capability of short vacuum gaps [4].

Prestrike in VIs happens when the movable contact is approaching the fixed contact during the current-making operation. Once the electric field strength across the contacts becomes larger than the breakdown strength of the vacuum gap, an electric arc may occur prior to the mechanical touch of the contacts, allowing a high inrush or making currents flow through. The prestrike phenomenon in VCBs can cause electromagnetic transients in electrical systems, which may lead to harmful effects and even damage in electrical devices in the system such as transformers, electrical machines and the circuit breaker itself [5]. The pre-breakdown phenomena in VCBs is complicated. There are mainly two theories about the pre-breakdown in VCBs. One is that field-emission current induces breakdown, the other is that microparticles induce vacuum breakdown [6]. If the field strength at the surface of a clean cathode in a VCB exceeds about 2×10^7 V/m, then field emission currents will be observed [7]. If the field strength exceeds about 1×10^8 V/m, pre-breakdown will occur [8]. Pre-breakdown induced by microparticles mainly occurs at larger gap spacing [9].

Numerous efforts have been made to understand the prestrike phenomenon in vacuum interrupters. Some researchers mainly focused on the melting effect of the prestrike arc on the contacts of the VCB, and this effect would affect the breaking performance of the VCB. To study this, Slade et al. [10,11] undertook a series of experiments to find that as the prestrike gap of the single-break VCB increased, the prestrike arc lasted longer. The prestrike arc could cause the contacts to weld and the damage caused by the welding depended on the arcing time. In order to determine the correlation of the prestrike process and the breaking process of the VCB which mainly was researched in the field of switching capacitor banks, Dullni et al. [12] found that the restrike phenomenon had some correlation with the prestrike process mainly due to the protrusions caused by the prestrike arc on the contacts in the single-break VCB. Körner et al. [13,14] obtained similar but more detailed results to Dullni. They found that the significant development of local protrusions on the contact surface affected both the microstructure and the macrostructure of contact surface during the capacitive making and breaking process. Some researchers focused on the welding force caused by the prestrike arc on the contacts of the VCB. If the value of the inrush current was high enough, the contacts would be melted and become welded. The welding force was a basic value that could reflect the melting degree of the contacts. Kumichev et al. [15] found that there was a correlation between welding forces caused by capacitive prestrike process and the number of non-sustained disruptive discharges (NSDDs). Yu et al. [16,17] found that both the amplitude and the scattering of the prestrike gaps and the welding force were significantly influenced by the contact materials about the capacitive current prestrike in single-break VCB. The damage caused by the prestrike arc was mainly influenced by the value of the prestrike current (5 kA, 10 kA or 20 kA). The material and structure of the contacts could also influence the prestrike arc behavior. Geng et al. [18] found that axial-magnetic field (AMF) contacts can effectively reduce the erosion of the contacts produced by the prestrike arc when the capacitive current is generated because of the diffusing effect produced by the AMF in single-break VCB, and higher inrush current value would cause larger damage on the contacts. Some researchers have undertaken calculations and simulations in order to gain an in depth understanding of the prestrike process. A.A. Razi-Kazemi et al. [19] established a new realistic transient model for prestrike phenomena in single-break VCB. Non-linear movement of the contacts, probabilistic behavior of the breakdown voltage and the non-linear dielectric strength curve of the VCB were fully considered in this model. Other researchers focused on the prestrike process mainly due to the design or application of the VCB. X. Ma et al. [20] undertook experiments to observe the direct current (DC) pre-breakdown and breakdown characteristics of micrometric gaps varying from 25 to 1000 μm. The results showed that the pre-breakdown conduction was dominated by high field-electron emission and could be valuable for the design of vacuum gaps in field emission displays. Kharin et al. [21,22], in a series of papers, undertook calculations and experiments to explore the complexity of closing single-break vacuum interrupter contacts, in order to make an electrical circuit. Sima et al. [23] found that phase-controlled VCBs can reduce the transient overvoltage and overcurrent compared with ordinary VCBs when they were used to switch 10 kV shunt capacitor banks. However, previous studies mainly focus on the prestrike characteristics of single-break VCBs. At present, experimental investigation of the prestrike characteristics of a double-break VCBs has not been reported so far.

The objective of this paper is to determine experimentally the probabilistic characteristics of prestrike gaps in a double-break VCB under DC voltages. This work can provide very important information on the prestrike arcing process, which is of significant importance for the study of the dielectric performance in VIs. Moreover, a better knowledge of the scatter in prestrike gaps in VIs can be very helpful to improve the control accuracy of phase-controlled switching.

2. Experimental Setup

Figure 1 shows an experimental circuit, including a capacitor bank C_s, a double-break VCB, a capacitive-resistance voltage divider, and an auxiliary electrical circuit. Figure 2 shows a picture of the experimental setup. During the experimental tests, the capacitor bank was initially charged to a certain voltage U_s, which was applied to the double-break VCB as soon as the switchgear SW_{DC} was closed. The VCB was then triggered to close the circuit. During the closing operation, the moving contact was approaching the fixed contact leading to a decrement of the dielectric strength of the vacuum gap between the contacts. When the pre-charged voltage exceeded the breakdown voltage of the vacuum gap, prestriking occurred before the mechanical touch of the contacts.

The double-break VCB under test included two commercial 12 kV vacuum interrupters in series, which were denoted VI_A and VI_B as shown in Figure 1. Cup-type AMF contacts were adopted. The contact material was CuCr30 (30% weight of Cr). The surfaces of the contacts were initially well-conditioned. A capacitive-resistance voltage divider, which can reduce the influence of the stray capacitance to ground, was used to measure the voltage U_m across the VCB during the transient prestrike process. When the first prestrike occurred, a stepdown in the measured voltage waveform of U_m was then captured. Moreover, with the help of an auxiliary circuit, which included a battery and two resistors in series that were connected to the auxiliary contact (SW_{au}) of the VCB, the time instant of the mechanical touch of the contacts in VIs can be detected from the waveform of voltage U_a. The displacement curve of the moving contact, denoted by L_{disp}, was measured by using a high-precision linear displacement transducer. Figure 3 shows an example of the waveforms of the measured voltages and contact displacement curve, where t_{pre} is the instant when the first prestrike occurs and t_{aux} is the instant when the moving contact is mechanically mated to the fixed contact. Finally, the prestrike gap d, i.e., the vacuum gap length between the contacts when the prestrike occurs during the closing operation, can be obtained.

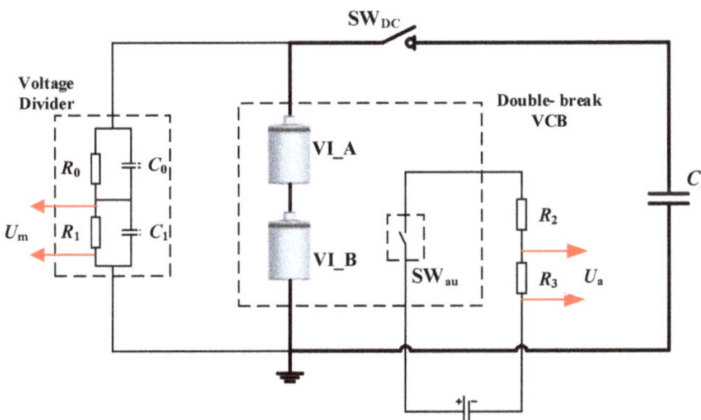

Figure 1. Schematic diagram of the experimental circuit.

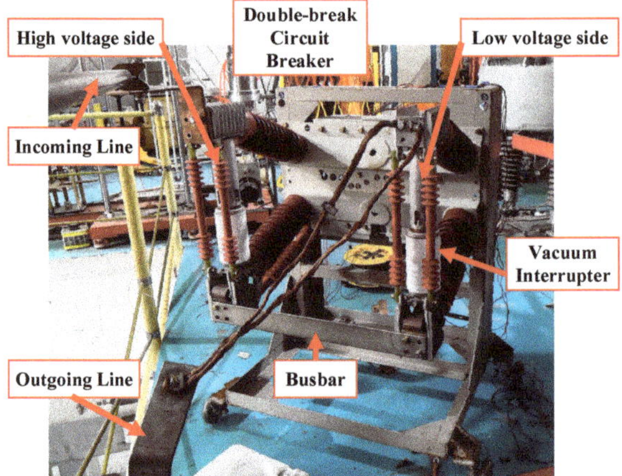

Figure 2. Picture of the experimental setup.

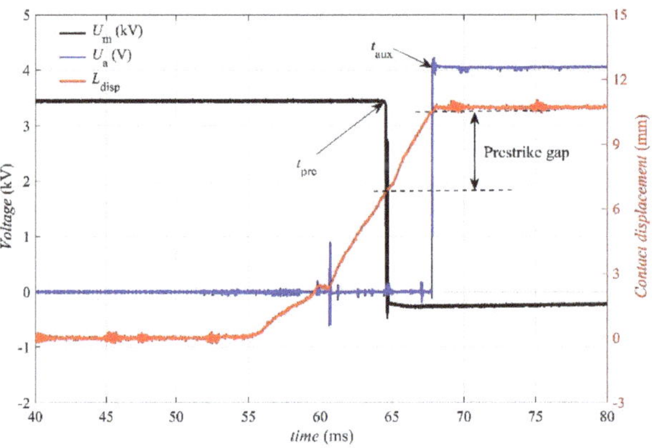

Figure 3. Example of the waveforms of voltages U_m and U_a and the moving.

Three groups of tests were carried out under the experimental conditions that are summarized in Table 1, where the applied DC voltage U_s were set as four levels of 10, 20, 30, 40 kV. In *Test* 1 and *Test* 2 the prestrike characteristics of vacuum interrupters (VI_A and VI_B) under DC voltages U_s were investigated separately. In this case, only one vacuum interrupter (VI_A or VI_B) was connected into the experimental circuit while the other one was shorted to ground by a busbar. In *Test* 3, the prestrike characteristics of the double-break VCB with two vacuum interrupters (VI_A together with VI_B) in series were analyzed, where VI_A was installed in the high-voltage side of the VCB and VI_B the low-voltage side (Figure 1). In this case, the total prestrike gap *d* was defined as the sum of the prestrike gaps in VI_A and in VI_B. The current-making operations were repeated 30 times in each test for all the test groups.

Table 1. Experimental conditions.

Test Group	Applied Voltage (kV)	Experimental Condition	Making Operation
Test 1	*Us*	Single-break test with VI_A	30
Test 2	*Us*	Single-break test with VI_B	30
Test 3	*Us*	Double-break test with VI_A and VI_B in series	30

The prestrike arc current during each current-making operation was only tens to hundreds of amperes. Considering all the contact surfaces had already been well conditioned, the erosion effect on the contact surfaces was low and could be neglected. Moreover, because of the asynchronous property of the mechanical actuators in the double-break VCB, the contacts in VI_A and VI_B were not mechanically closed at the same time. The average time delay of VI_B compared with VI_A was 0.1 ms (computed from 50 closing operations with no load), which was used as a compensation when calculating the time instant t_{aux}.

3. Experimental Results

3.1. Probabilistic Characteristics of the Pestrike Gap

The prestrike phenomenon in the vacuum circuit breakers can be characterized by prestrike gap d, which is the instantaneous vacuum gap length between the contacts of the vacuum interrupters. It is measured at the occurrence of the first prestrike during a current-making operation. In this work, a probabilistic analysis of the prestrike gap under different applied voltage levels had been carried out. The distributions of the prestrike gaps in the test groups under each voltage level were calculated. Let $f(d)$ be the density function (DF) of the prestrike gap distributions, which indicates the chance of the prestrike between the contacts in VIs at a vacuum gap length d. The corresponding complementary cumulative distribution function (CCDF) can be then calculated by:

$$F(d) = \int_{d}^{\infty} f(u)du \tag{1}$$

which gives the probability of the prestrike between the contacts in VIs when the vacuum gap length is no less than d. Figures 4–6 show the calculated CCDF under applied voltage U_s from the experimentally measured prestrike gaps in *Test* 1, *Test* 2 and *Test* 3, respectively. In *Test* 3, the total prestrike gap as well as the prestrike gaps in VI_A (high voltage side) and in VI_B (low voltage side) were all given.

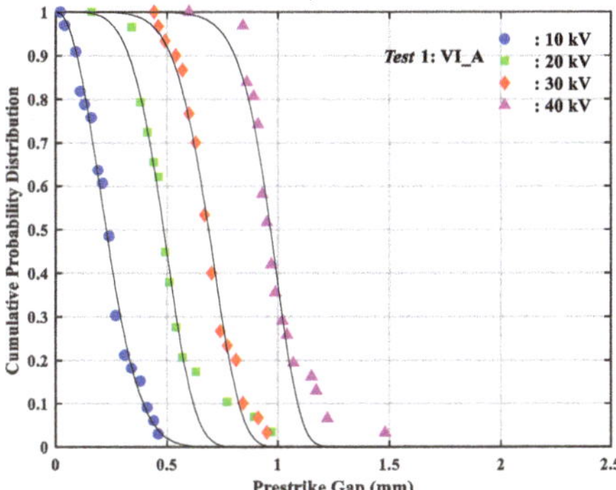

Figure 4. The complementary cumulative distribution function of prestrike gaps in *Test* 1.

For the characterization of the CCDF, the Weibull distribution is used and the CCDF is then given by:

$$F(d) = \exp\left(-\left(\frac{d}{\eta}\right)^{\beta}\right) \tag{2}$$

where $\eta > 0$ is the shape parameter of the Weibull distribution [24] and $\beta > 0$ is the scale parameter of the Weibull distribution or the characteristic value of the prestrike gap. Both η and β are obtained by fitting the experimental data, as shown in Table 2.

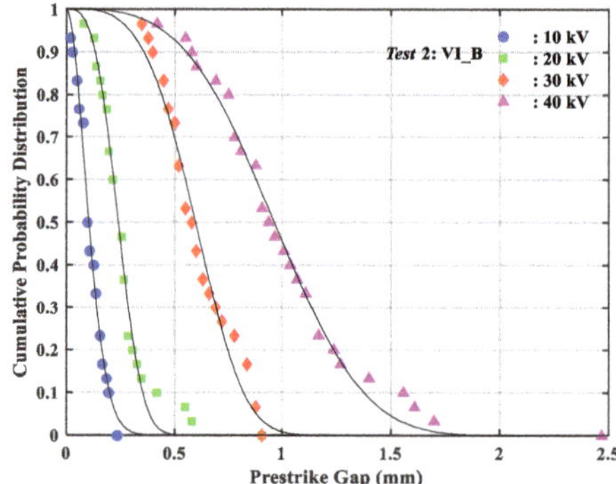

Figure 5. The complementary cumulative distribution function of prestrike gaps in *Test* 2.

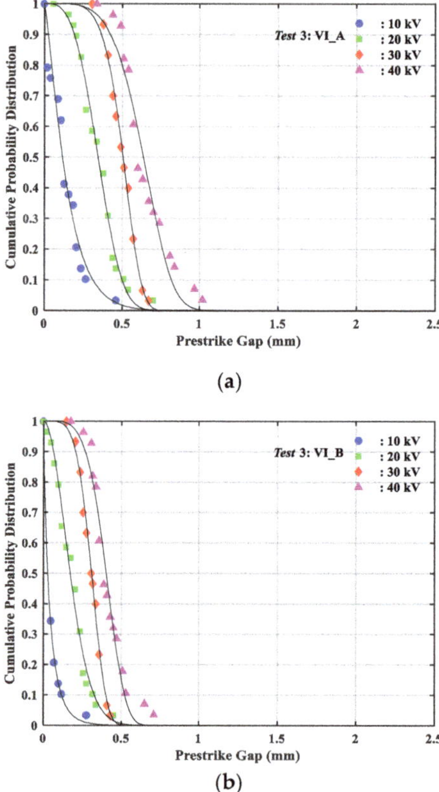

Figure 6. *Cont.*

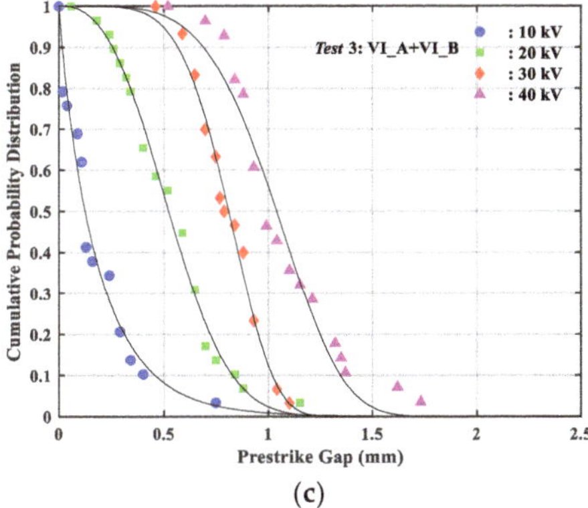

(c)

Figure 6. The complementary cumulative distribution function of prestrike gaps in *Test* 3 with VI_A and VI_B in series: complementary cumulative distribution function (CCDF) of prestrike gaps in VI_A (**a**), in VI_B (**b**), and CCDF of the total prestrike gaps (**c**).

Table 2. Values of the Weibull parameters.

Test Group		*Us* (kV)	β	η (mm)	R-Square
Test 1 (VI_A)		10	2.2	0.3	0.99
		20	4.9	0.5	0.98
		30	6.7	0.7	0.99
		40	10.3	1.0	0.95
Test 2 (VI_B)		10	1.8	0.1	0.99
		20	3.2	0.3	0.99
		30	3.7	0.7	0.97
		40	3.4	1.1	0.99
Test 3	VI_A and VI_B	10	0.9	0.2	0.97
		20	2.6	0.6	0.99
		30	5.3	0.9	0.99
		40	4.6	1.1	0.97
	VI_A	10	1.2	0.2	0.95
		20	3.3	0.4	0.99
		30	5.9	0.5	0.99
		40	4.7	0.7	0.96
	VI_B	10	0.8	0.04	0.99
		20	1.9	0.2	0.99
		30	4.7	0.3	0.99
		40	4.7	0.4	0.97

The fitted curves with the Weibull distribution given in (2) with respect to *Test* 1, *Test* 2 and *Test* 3 are shown in Figures 4–6, respectively, where a good agreement between the fitted curves and the experimental data can be observed. Besides the evaluation of the fit from the figures graphically, R-square is used in order to evaluate the goodness of the fit numerically, which is the square of the correlation between the predicted prestrike gaps and the experimentally measured values. R-square can take on any value between zero and one, where a value closer to one indicates a better performance of the fitted model [24–26]. The calculated values of the R-square parameter are presented in the last column of Table 2. From Table 2 it can be noted that the values of R-square are no smaller than 95% in all tests, which indicates that the prestrike gap has a Weibull distribution with parameters η and β

and that the model with the given parameters fits the prestrike gap distributions well. CCDF density function $f(d)$ is given by

$$f(d|\eta,\beta) = \frac{\beta}{\eta}\left(\frac{d}{\eta}\right)^{\beta-1} \exp\left(-\left(\frac{d}{\eta}\right)^{\beta}\right) \tag{3}$$

3.2. Scatters in the Prestrike Gap

The scatter in the prestrike gap is of significant importance for investigating the prestrike phenomenon in vacuum interrupters. Moreover, for the phase-controlled switching application with vacuum circuit breakers, a better knowledge of the scatter in the prestrike gap could help to improve control accuracy. The scatter in the prestrike gap can be characterized by the standard deviation of the experimental measurement of the prestrike gaps, which is given by:

$$\sigma = \sqrt{\frac{1}{N}\sum_{i=1}^{N}\left(d_i - \overline{d}\right)^2} \tag{4}$$

where N is the number of each experimental data set; $\overline{d}$ is the average value of each set of the experimental data, and d_i is the prestrike gap measured during the i_{th} making operation. The scatters in the prestrike gaps in all the test groups under varied applied voltages are given in Table 3. In the last column, σ indicates the scatter, where a lager value of σ indicates more scatter in the experimental data of the prestrike gaps.

Table 3. Prestrike gaps of 10%, 50% and 90% and scatters in prestrike gaps.

Test Group		Us (kV)	d_{10} (mm)	d_{50} (mm)	d_{90} (mm)	σ (mm)
Test 1 (VI_A)		10	0.39	0.23	0.10	0.11
		20	0.62	0.48	0.33	0.16
		30	0.82	0.69	0.52	0.12
		40	1.09	0.97	0.81	0.15
Test 2 (VI_B)		10	0.20	0.11	0.04	0.06
		20	0.35	0.24	0.13	0.16
		30	0.82	0.60	0.36	0.17
		40	1.38	0.96	0.55	0.41
Test 3	VI_A and VI_B	10	0.46	0.13	0.02	0.16
		20	0.83	0.52	0.25	0.23
		30	1.02	0.81	0.57	0.16
		40	1.35	1.04	0.69	0.27
	VI_A	10	0.32	0.12	0.03	0.11
		20	0.50	0.35	0.19	0.13
		30	0.62	0.50	0.37	0.09
		40	0.82	0.63	0.42	0.16
	VI_B	10	0.12	0.03	0.00	0.06
		20	0.33	0.18	0.07	0.10
		30	0.40	0.31	0.21	0.07
		40	0.52	0.40	0.27	0.11

To study the probabilistic characteristics of prestrike gaps, the 10% prestrike gap d_{10}, 50% prestrike gap d_{50}, and 90% prestrike gap d_{90}, i.e., the prestrike gap at which the value of CCDF is 10%, 50%, and 90%, respectively, are commonly used [7,16,25,26]. In this case, the influence of the approximation error between the fitted CCDF and the measured prestrike gaps that are higher than d_{90} or lower than d_{10} on the probabilistic characteristics of prestrike gaps can be neglected. The values of d_{10}, d_{50}, and d_{90} can be calculated by (2), which are shown in Table 3. It can be noted that with the increment of applied voltage U_s, the d_{10}, d_{50} and d_{90} are all increasing significantly. Moreover, the 50% prestrike gap d_{50} can be used as one of the most important parameters since it gives the prestrike gap at which there is a

50% chance for the vacuum gap to achieve breakdown in the VCB, i.e., d_{50} acts as the mean value of the prestrike gap in the sense of the Weibull distribution. The relationships between d_{50} and applied voltage U_s in *Test* 1, *Test* 2, and *Test* 3 are shown in Figure 7, Figure 8, and Figure 9, respectively, where the error bar at each data point shows the corresponding scatter in the prestrike gaps. It can be noted that the value of d_{50} is approximately proportional to applied voltage U_s.

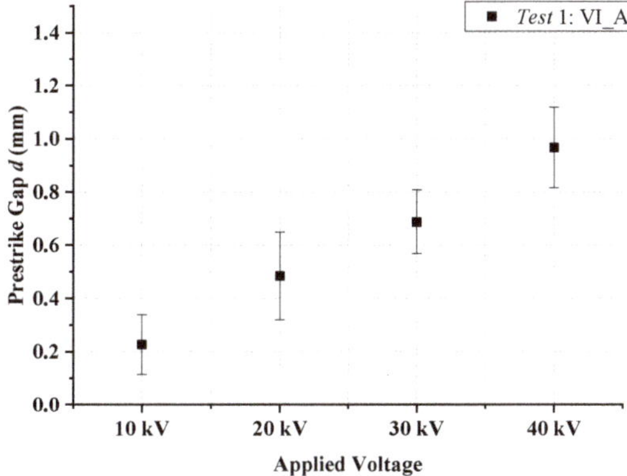

Figure 7. The relationships between d_{50} and applied voltage *Us* in *Test* 1 with VI_A.

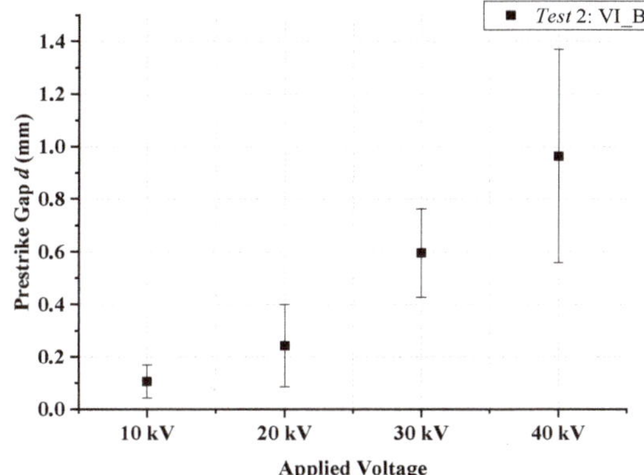

Figure 8. The relationships between d_{50} and applied voltage *Us* in *Test* 2 with VI_B.

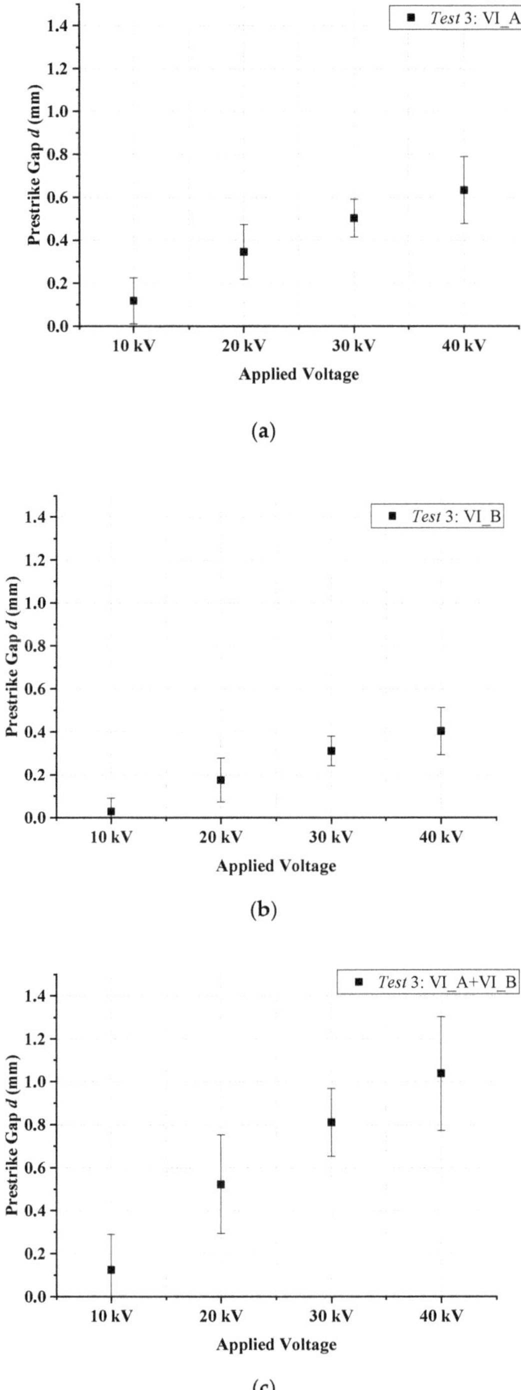

Figure 9. The relationships between d_{50} and applied voltage *Us* in *Test* 3: VI_A (**a**), VI_B (**b**), and the total prestrike gaps (**c**).

4. Discussion

4.1. Positive Effects of the Double-Break Vacuum Circuit Breaker

A vacuum interrupter in the double-break tests (*Test* 3) may have different prestrike characteristics in terms of prestrike gaps and scatters in prestrike gaps compared with that in the single-break tests (*Test* 1 and *Test* 2) due to the voltage-sharing-effect, the asynchronous property of the mechanical actuators, and the inhomogeneity of the inherent prestrike characteristics of VIs in a double-break VCB.

From Table 3, it can be noted that the scatters in the prestrike gaps of VI_A (resp., VI_B) in the double-break tests, i.e., in *Test* 3 are smaller than that of VI_A (VI_B) in the single-break tests, i.e., in *Test* 1 (with respect to *Test* 2). Vacuum interrupters with less scatters in the prestrike gaps would be of great importance in order to improve the accuracy of the phase-controlled making technique. Therefore, with the same VIs, a double-break VCB would be better suitable for the phase-controlled switching.

The double-break VCB takes full advantage of the good dielectric insulation ability of the vacuum gap. A double-break VCB which has two gaps in series can withstand higher voltage compared to the single-break VCB in the same total break length [27]. The relationship between breakdown voltage and electrodes gap of the single-break VCB has two different expressions. When the electrode gap is less than 5 mm, the breakdown voltage and electrodes gap length is a linear equation, as shown in:

$$U_t = k \cdot d \tag{5}$$

When the electrode gap is greater than 5 mm, Breakdown voltage and electrodes gap length is a non-linear equation when the electrodes' distance is greater than 5 mm, as shown in:

$$U_t = k \cdot d^m \tag{6}$$

The value of m is between 0.4 to 0.7, d stands for the electrodes' gap length, k is a constant number.

The relationship between breakdown voltage and electrodes' gap length of the double-break VCB is shown in (7). It is based on the assumption that the voltage distribution value of two interrupters is equal.

$$U_t = 2k \cdot d^m \tag{7}$$

However, in the double-break VCB which does not have a grading capacitor in parallel with each VI, the shared voltages of each VI are not equal because of the stray capacitance. It is impossible to realize double withstand voltage as (7) shows [28]. In order to investigate the influence of the voltage-sharing effect in double-break VCBs on the prestrike characteristics, the relative variations in the prestrike gaps of each vacuum interrupter under different applied voltage levels are calculated by:

$$\delta(d) = \frac{d^{(DB)} - d^{(SB)}}{d^{(SB)}} \times 100\% \tag{8}$$

where $d^{(DB)}$ is the prestrike gap of vacuum interrupter VI_A (resp., VI_B) obtained from the double-break test, i.e., *Test* 3 and $d^{(SB)}$ is the prestrike gap from the single-break test, i.e., *Test* 1 (resp., *Test* 2).

Table 4 shows the values of relative variations δ in d_{10}, d_{50} and d_{90}. The vacuum gaps in VI_A and VI_B in the double-break test can be considered as two capacitors in series [29] and each gap shared part of the total applied voltage U_s. Therefore, the voltage applied to each gap was smaller than U_s, leading to a smaller prestrike gap. The smaller the prestrike gap was, the shorter the prestrike arcing time was. For a given applied voltage, a double-break VCB has a better performance than a single-break one in terms of reducing the prestrike gaps in the vacuum interrupter, which is termed as the positive effects of double-break VCBs.

Table 4. Relative variations in the 10%, 50% and 90% prestrike gaps.

Us (kV)		10	20	30	40
VI_A	δ (d_{10})	−17.9%	−19.4%	−24.4%	−24.8%
	δ (d_{50})	−47.8%	−27.1%	−27.5%	−35.1%
	δ (d_{90})	−70%	−42.4%	−28.8%	−48.1%
VI_B	δ (d_{10})	−40%	−5.7%	−51.2%	−62.3%
	δ (d_{50})	−72.7%	−25.0%	−48.3%	−58.3%
	δ (d_{90})	−100%	−46.2%	−41.7%	−50.9%

4.2. Negative Effects of the Double-Break Vacuum Circuit Breaker

Due to the voltage-sharing effect, the actual voltage applied to each vacuum interrupter VI_A or VI_B in the double-break tests was smaller than that in the single-break tests. In order to investigate the prestrike characteristics of vacuum interrupters in double-break tests by comparison with that in single-break tests under the same applied voltage, the influence of the voltage-sharing effect has to be eliminated.

In an ideal case, if both interrupters VI_A and VI_B share the same percentile of the total applied voltage U_s, i.e., both share 50% out of U_s, then the sharing voltage of VI_A (resp., VI_B) in *Test* 3 would be 10 kV or 20 kV, i.e., the same voltage applied to VI_A (resp., VI_B) in *Test* 1 (resp., *Test* 2), when the applied voltage to the double-break U_s in *Test* 3 is 20 kV or 40 kV. Figure 10 shows the 50% prestrike gap d_{50} of VI_A and VI_B in the single-break tests compared with that in the double-break tests where the value of applied voltage U_s in *Test* 3 is twice as large as that in single-break tests (*Test* 1 and *Test* 2).

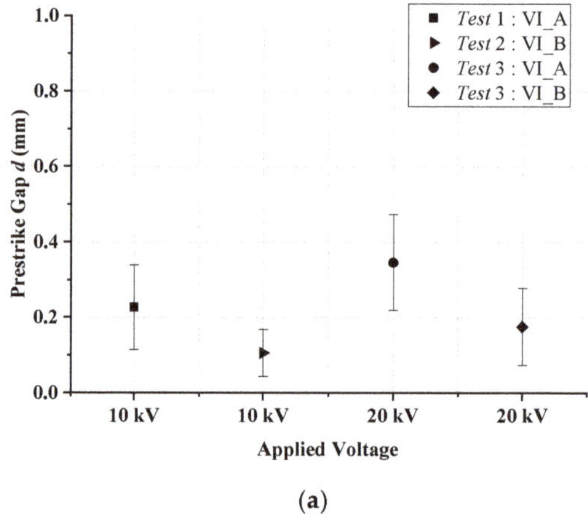

(a)

Figure 10. *Cont.*

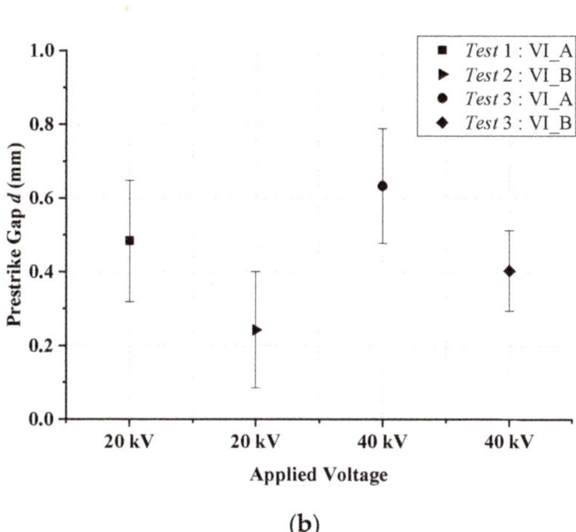

(b)

Figure 10. The negative effects of double-break vacuum circuit breakers (VCBs) on the prestrike gaps of vacuum interrupters. (**a**) The single-break VCB under 10 kV compared with the double-break under 20kV. (**b**) The single-break VCB under 20 kV compared with the double-break under 40kV.

However, the voltage-sharing ratio of VI_A and VI_B in double-break VCBs was not the same due to the stray capacitance to ground. In fact, from the previous studies the voltage-sharing ratio of the vacuum interrupter installed on the high voltage side (i.e., VI_A in our case) is larger than that on the low voltage side (VI_B) [30]. Note that a larger applied voltage leads to a higher prestrike gap. Therefore, the prestrike gaps of VI_A in *Test* 3 should be larger than that in *Test* 1, while the prestrike gaps of VI_B in *Test* 3 should be smaller than that in *Test* 2 when U_s in *Test* 3 is twice as large as that in *Test* 1 and *Test* 2. This can be verified for the vacuum interrupter VI_A from Figure 10, but not for VI_B. As can be observed from Figure 10, the prestrike gap of VI_B in *Test* 3 is larger. To be precise, the voltage applied to VI_B in *Test* 3 with U_s = 20 kV (resp., 40 kV) is lower than 10 kV (resp., 20 kV) due to the voltage-sharing effect while the 50% prestrike gaps d_{50} is larger than that in *Test* 2 with an applied voltage of 10 kV (with respect to 20 kV) instead. That is to say, the double-break VCB does not make fully advantage of the dielectric strength of the vacuum gap on the low voltage side, which is termed as the negative effects of double-break VCBs.

The negative effects of double-break VCBs may be caused by the asynchronous property of the mechanical actuators, the inhomogeneity of the inherent prestrike characteristics and the unequal voltage-sharing ratio of VIs in a double-break VCB.

The double-break VCB is closed or interrupted by the mechanical actuators. In the ideal case, the two VIs of the double-break VCB moves at the same velocity and the operating characteristics are same too. Due to the inevitable performance differences between the mechanism's actuators of the double-break VCB, and their tolerance to various environmental changes not being the same, the effect caused by the asynchronous property of the mechanical actuators on the double-break VCB must not be underestimated. In the breaking process, when the operating time interval of two breaks is 2 ms, the distance difference between two vacuum gaps is significant. The arcing time of the two VIs is different, which makes the sharing voltage of each VI is unequal in the breaking process. In the most serious situation, reignition will occur [31]. On the other hand, in the current-making process, considering a double-break VCB with two vacuum interrupters VI_A and VI_B in series, we denote by $d^*_{50}(\text{VI\_A})$ and $d^*_{50}(\text{VI\_B})$ the inherent 50% prestrike gaps of VI_A and VI_B, which is the 50% prestrike gap when tested separately in the single-break tests. Due to the asynchronous property of the mechanical actuators in the double-break VCB, one of the moving contacts, for example the moving contact in VI_A, moves faster than the other, as shown in Figure 11a. Let both VI_A and VI_B share half of the applied voltage ($U_s/2$) and have a same inherent 50% prestrike gap, i.e., $d^*_{50}(\text{VI\_A}) = d^*_{50}(\text{VI\_B})$.

During the closing operation, the vacuum gap length between the contacts in VI_A is smaller than that in VI_B ($l_A < l_B$). Therefore, it is very possible for VI_A to breakdown earlier than VI_B. As soon as the prestrike occurs in VI_A, the total voltage U_s would be applied to the vacuum gap in VI_B. Then the prestrike occurs in VI_B. In this case, the measured prestrike gap of VI_B is the same as l_A, which is larger than $d^*_{50}(\text{VI\_B})$ the inherent 50% prestrike gap of VI_B.

Moreover, let both VI_A and VI_B be synchronized perfectly and share half of the applied voltage ($U_s/2$), as shown in Figure 11b. The inherent 50% prestrike gap of VI_A is larger than that of VI_B, i.e., $d^*_{50}(\text{VI\_A}) > d^*_{50}(\text{VI\_B})$. Then, it is very possible that the prestrike occurs in VI_A earlier that in VI_B, leading to a higher prestrike gap of VI_B (that is l_B) than $d^*_{50}(\text{VI\_B})$.

Similar effects can also be found in the case when the voltage-sharing ratios of VI_A and VI_B are not equal to each other. As shown in Figure 11c, both VI_A and VI_B are synchronized perfectly and have the same inherent 50% prestrike gap, since VI_A on the high voltage side shares voltage higher than $U_s/2$. The interrupter VI_A is highly possible to breakdown when the vacuum gap length is larger than the inherent 50% prestrike gap, i.e., $l_A > d^*_{50}(\text{VI\_A})$. Noting VI_A and VI_B are perfectly synchronized ($l_A = l_B$), the measured 50% prestrike gap of VI_B would be larger than its inherent one.

The mechanical actuators of the double-break VCB used in this work are not perfectly synchronized, which is practically impossible. The mechanical actuator installed on the low-voltage side moves slower than that on the high voltage side with a time delay of 0.1 ms. Moreover, from Table 3, the inherent 50% prestrike gap of VI_A is always larger than that of VI_B under the tested applied voltages. Together with the effect of the unequal voltage-sharing ratio, the negative effects of double-break VCBs are observed as shown in Figure 10.

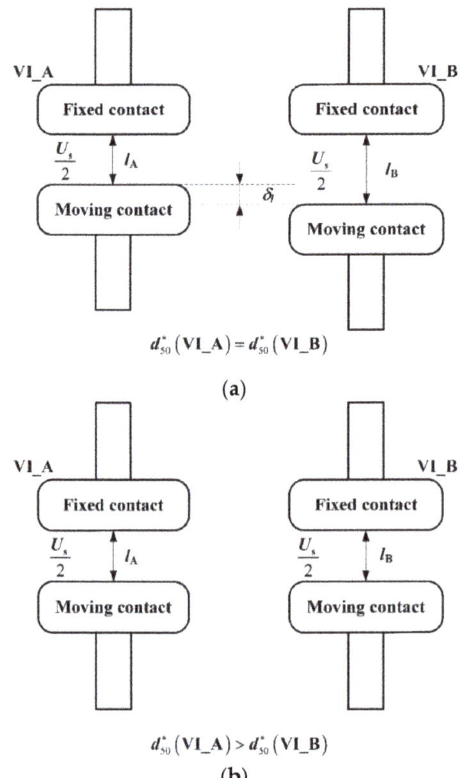

Figure 11. *Cont.*

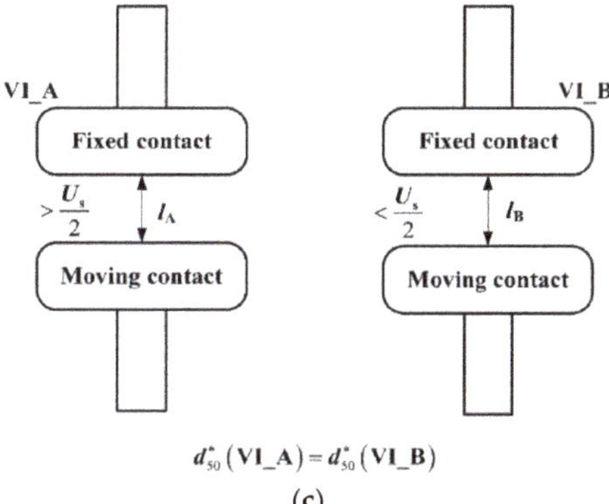

Figure 11. Schematic diagram of the negative effects of the double-break VCBs in terms of the asynchronous property of the mechanical actuators (**a**), the inhomogeneity of the inherent prestrike gaps (**b**), and the unequal voltage-sharing ratio (**c**).

The prestriking phenomena are irrelevant to the initial contact gaps in the double break and the single break in this work. As for the experimental setup, the double-break VCB consisted of two VIs. When carrying out the single-break tests, only one vacuum interrupter (VI_A or VI_B) was connected into the experimental circuit while the other one was shorted to ground by the busbar. The initial gaps of both VIs were 10 mm. Therefore, the initial gap of the single-break VCBs was 10 mm, and the initial gap of the double-break VCB was 20 mm (two 10 mm gaps in series). There are mainly two theories about the breakdown in the vacuum. One is that field emission induces breakdown, the other is that particles induce vacuum breakdown [6]. If the vacuum gap is less than 2 mm, the field emission plays the dominant role in the breakdown process [9]. In these tests, all the prestrike gaps were less than 2 mm. The prestrike was field-emission induced. The prestrike arc current during each making operation was only tens to hundreds of amperes, the erosion effect on the contact surfaces can be neglected. During the making process of the VCB, there is a threshold electric field strength that determined whether breakdown occurs or not [32]. When the electric field strength exceeds the threshold electric field strength between the contacts, the prestrike arc occurs and the gap between the contacts at that instant is called prestrike gap, which is independent of the initial gap of the VCB [33]. However, if there is an inrush current in the prestrike process that causes erosion on contacts or the prestrike gaps are larger than 2 mm, the initial gap of the single-break VCB and the double-break VCB must be kept the same in order to compare the characteristics of prestrike gaps.

5. Conclusions

In this paper, the prestrike characteristics of a double-break VCB consisting of two 12 kV VIs in series under DC voltages had been experimentally analyzed. The following conclusions can be drawn.

(1) With the increment of the applied DC voltage, the 10% prestrike gap d_{10}, 50% prestrike gap d_{50} and 90% prestrike gap d_{90} were all increasing significantly, whereas the scatters in the prestrike gaps were not changing too much. Specifically, the value of d_{50} was approximately proportional to the applied voltage.

(2) With a given applied DC voltage, the positive effect of the double-break VCBs on the prestrike characteristics can be observed, i.e., the prestrike gaps of the vacuum interrupters in the double-break tests can be significantly reduced in comparison with that in the single-break tests due to the voltage-sharing effect in double-break VCBs. Moreover, fewer scatters in the prestrike gaps during the double-break tests can be found.

(3) The double-break VCB with two vacuum interrupters in series did not take full advantage of the dielectric strength of the vacuum gap on the low-voltage side, i.e., the negative effect of a

double-break VCB on the prestrike characteristics, which may be caused by the asynchronous property of the mechanical actuators, the inhomogeneity of the inherent prestrike characteristics and the unequal voltage-sharing ratio of VIs in a double-break VCB.

Author Contributions: Y.G. (Yun Geng) and X.Y. contributed to the research idea and designed the experiments. Y.G. (Yun Geng) and J.D. worked on the data analysis and drafted the manuscript. X.L. and J.P. helped to do the experiments. Y.G. (Yingsan Geng), Z.L. and K.W. worked on editing and revising of the manuscript. All authors have read and approved the final manuscript.

Funding: This work was supported in part by the National Natural Science Foundation of China under Grant 51937009, Key Research and Development Program of Shaanxi (2019ZDLGY18-04), the Fundamental Research Funds for the Central Universities (xzy022019013).

Conflicts of Interest: The authors declare no conflict of interest.

References

1. Ide, N.; Tanaka, O.; Yanabu, S.; Kaneko, S.; Okabe, S.; Matsui, Y. Interruption characteristics of double-break vacuum circuit breakers. *IEEE Trans. Dielectr. Electr. Insul.* **2008**, *15*, 1065–1072. [CrossRef]
2. Ge, G.; Liao, M.; Duan, X.; Cheng, X.; Zhao, Y.; Liu, Z.; Zou, J. Experimental Investigation into the Synergy of Vacuum Circuit Breaker with Double-Break. *IEEE Trans. Plasma Sci.* **2016**, *44*, 79–84. [CrossRef]
3. Giere, S.; Karner, H.C.; Knobloch, H. Dielectric strength of double and single-break vacuum interrupters: Experiments with real HV demonstration bottles. *IEEE Trans. Dielectr. Electr. Insul.* **2001**, *8*, 43–47. [CrossRef]
4. Fugel, T.; Koenig, D. Switching performance of two 24 kV vacuum interrupters in series. *IEEE Trans. Dielectr. Electr. Insul.* **2002**, *9*, 164–168. [CrossRef]
5. Delachaux, T.; Rager, F.; Gentsch, D. Study of vacuum circuit breaker performance and weld formation for different drive closing speeds for switching capacitive current. In Proceedings of the 24th International Symposium on Discharges and Electrical Insulation in Vacuum (ISDEIV2010), Braunschweig, Germany, 30 August–3 September 2010; pp. 241–244.
6. Zhang, Y.; Xu, X.; Jin, L.; An, Z.; Zhang, Y. Fractal-based electric field enhancement modeling of vacuum gap electrodes. *IEEE Trans. Dielectr. Electr. Insul.* **2017**, *24*, 1957–1964. [CrossRef]
7. Slade, P.G. *The Vacuum Interrupter Theory, Design and Application*; CRC Press: Boca Raton, FL, USA, 2008.
8. Descoeudres, A.; Ramsvik, T.; Calatroni, S.; Taborelli, M.; Wuensch, W. DC breakdown conditioning and breakdown rate of metals and metallic alloys under ultrahigh vacuum. *Phys. Rev. Spec. Top. Accel. Beams* **2009**, *12*. [CrossRef]
9. Latham, R.V. *High Voltage and Vacuum Insulation: Basic Concepts and Technological Practice*; Academic Press: London, UK, 1995.
10. Slade, P.G.; Taylor, E.D.; Haskins, R.E. Effect of short circuit current duration on the welding of closed contacts in vacuum. In Proceedings of the 51th IEEE Holm Conference on Electrical Contacts, Chicago, IL, USA, 26–28 September 2005; pp. 69–74.
11. Slade, P.G.; Smith, R.K.; Taylor, E.D. The effect of contact closure in vacuum with fault current on prestrike arcing time, contact welding and the field enhancement factor. In Proceedings of the 2007 53rd IEEE Holm Conference on Electrical Contacts, Pittsburgh, PA, USA, 16–19 September 2007; pp. 32–36.
12. Dullni, E.; Shang, W.; Gentsch, D.; Kleberg, I.; Niayesh, K. Switching of Capacitive Currents and the Correlation of Restrike and Pre-ignition Behavior. *IEEE Trans. Dielectr. Electr. Insul.* **2006**, *13*, 65–71. [CrossRef]
13. Körner, F.; Lindmayer, M.; Kurrat, M.; Gentsch, D. Contact Behavior in vacuum under capacitive switching duty. *IEEE Trans. Dielectr. Electr. Insul.* **2007**, *14*, 643–648. [CrossRef]
14. Körner, F.; Lindmayer, M.; Kurrat, M.; Gentsch, D. Switching behavior of different contacts materials under capacitive switching conditions. In Proceedings of the 23rd International Symposium on Discharges Electrical Insulation in Vacuum, Braunschweig, Germany, 15–19 September 2008; pp. 202–205.
15. Kumichev, G.A.; Poluyanova, I.N. Investigation of welding characteristics and NSDD probabilities of different contact materials under capacitive load conditions. In Proceedings of the 27th International Symposium on Discharges and Electrical Insulation in Vacuum (ISDEIV), Suzhou, China, 18–23 September 2016; pp. 1–4.

16. Yu, Y.; Li, G.; Geng, Y.; Wang, J.; Liu, Z. Prestrike inrush current arc behaviors in vacuum interrupters subjected to a transverse magnetic field and an axial magnetic field. *IEEE Trans. Plasma Sci.* **2018**, *46*, 3075–3082. [CrossRef]
17. Yu, Y.; Geng, Y.; Geng, Y.; Wang, J.; Liu, Z. Inrush current prestrike arc behaviours of contact materials CuCr50/50 and CuW10/90. *IEEE Trans. Plasma Sci.* **2017**, *45*, 266–274. [CrossRef]
18. Geng, Y.; Yu, Y.; Geng, Y.; Wang, J.; Liu, Z. Inrush current arc characteristics in vacuum interrupters with axial magnetic field contact and butt-type contact. In Proceedings of the 3rd International Conference on Electric Power Equipment—Switching Technology (ICEPE-ST2015), Busan, Korea, 25–28 October 2015; pp. 512–515.
19. Razi-Kazemi, A.A.; Fallah, M.R.; Rostami, M.; Malekipour, F. A new realistic transient model for restrike/prestrike phenomena in vacuum circuit breaker. *Int. J. Electr. Power Energy Syst.* **2020**, *117*. [CrossRef]
20. Ma, X.; McLester, M.; Sudarshan, T.S. Prebreakdown and breakdown characteristics of micrometric vacuum gaps between broad area electrodes. In Proceedings of the IEEE 1997 Annual Report Conference on Electrical Insulation and Dielectric Phenomena, Minneapolis, MN, USA, 19–22 October 1997; pp. 583–586.
21. Kharin, S.N.; Nouri, H.; Amft, D. Dynamics of electrical contact floating in vacuum. In Proceedings of the 48th IEEE Holm Conference on Electrical Contacts, Orlando, FL, USA, 22–23 October 2002; pp. 197–205.
22. Kharin, S.N.; Ghori, Q.K. Influence of the pre-arcing bridging on the duration of vacuum arc. In Proceedings of the 19th International Symposium on Discharges and Electrical Insulation in Vacuum, Xi'an, China, 18–22 September 2000; pp. 278–285.
23. Sima, W.; Zou, M.; Yang, Q.; Yang, M.; Li, L. Field experiments on 10 kV switching shunt capacitor banks using ordinary and phase-controlled vacuum circuit breakers. *Energies* **2016**, *9*, 88. [CrossRef]
24. Mccool, J.I. *Using the Weibull Distribution Reliability: Modeling, and Inference*; Wiley & Sons Inc.: Hoboken, NJ, USA, 2012.
25. Zhang, Y.; Yang, H.; Wang, J.; Geng, Y.; Liu, Z.; Jin, L.; Yu, L. Influence of high-frequency high-voltage impulse conditioning on back-to-back capacitor bank switching performance of vacuum interrupters. *IEEE Trans. Plasma Sci.* **2016**, *44*, 321–330. [CrossRef]
26. Yang, H.; Geng, Y.; Liu, Z.; Zhang, Y.; Wang, J. Capacitive switching of vacuum interrupters and inrush currents. *IEEE Trans. Dielectr. Electr. Insul.* **2014**, *21*, 159–170. [CrossRef]
27. Cheng, X.; Liao, M.; Duan, X.; Zou, J. Experimental research on dynamic dielectric recovery characteristics for vacuum switch with double-break. In Proceedings of the 2011 IEEE Pulsed Power Conference, Chicago, IL, USA, 19–23 June 2011; pp. 291–296.
28. Shiba, Y.; Ide, N.; Ichikawa, H.; Matsui, Y.; Sakaki, M.; Yanabu, S. Withstand Voltage Characteristics of Two Series Vacuum Interrupters. *IEEE Trans. Plasma Sci.* **2007**, *35*, 879–884. [CrossRef]
29. Liao, M.; Duan, X.; Cheng, X.; Zou, J. Dynamic dielectric recovery property for vacuum circuit-breakers with double breaks. In Proceedings of the 24th International Symposium on Discharges and Electrical Insulation in Vacuum (ISDEIV2010), Braunschweig, Germany, 30 August–3 September 2010; pp. 225–228.
30. Fugel, T.; Koenig, D. Breaking performance of a capacitive-graded series design of two 24-kV vacuum circuit breakers. In Proceedings of the 20th International Symposium on Discharges and Electrical Insulation in Vacuum (ISDEIV2002), Tours, France, 1–5 July 2002; pp. 360–363.
31. Betz, T.; Konig, D. Influence of grading capacitors on the breaking capacity of two vacuum interrupters in series. *IEEE Trans. Dielectr. Electr. Insul.* **1999**, *6*, 405–409. [CrossRef]
32. Li, S.; Geng, Y.; Liu, Z.; Wang, J. A breakdown mechanism transition with increasing vacuum gaps. *IEEE Trans. Dielectr. Electr. Insul.* **2017**, *24*, 3340–3346.
33. Wei, L.; Chun-En, F.; Bi-De, Z.; Pian, X.; Xiao, R.; Yan, L. Research on controlled switching in reducing unloaded power transformer inrush current considering circuit breaker's prestrike characteristics. In Proceedings of the 27th International Symposium on Discharges and Electrical Insulation in Vacuum (ISDEIV 2016), Suzhou, China, 18–23 September 2016; pp. 1–4.